CU00767657

David McLaren

Spreadsheets and Numerical Analysis

 Chartwell-Bratt    Studentlitteratur

All rights reserved. No part of this publication may be reproduced or
transmitted in any form or by any means, electronic or mechanical,
including photocopying, recording, or any information storage and retrieval system,
without permission in writing from the publisher.

© David McLaren and Chartwell-Bratt 1997.

Chartwell-Bratt (Publishing and Training) Ltd
ISBN 0-86238-431-1

Printed in Sweden
Studentlitteratur, Lund
Art.no. 6621
ISBN 91-44-00582-2

Printing/Year		1	2	3	4	5	6	7	8	9	10	2001	2000	99	98	97

CONTENTS

Preface

The aim of this book is to introduce the reader to some aspects of the numerical solution of mathematical problems. The intention is that the book should provide a rapid and easy "hands on" first encounter with the subject. It emphasizes learning by doing, using the spreadsheet as a convenient platform.

Using a spreadsheet has the advantage that the user need not first acquire a working knowledge of a mathematically oriented computer language such as FORTRAN or Pascal before embarking on the exploration of numerical techniques. An additional advantage is that a spreadsheet makes the details of the calculations and any resulting graphical displays simultaneously visible. Further, the effects of parameter changes are rapidly recalculated and graphs promptly updated.

There are many types of problem that can be solved on a spreadsheet, and many that are impossible or highly inconvenient to attempt to solve this way. For problems that are amenable to spreadsheet solution, the main effort required is some forethought and planning. This commitment would in any case be needed for the use of an alternative computing medium.

The text includes examples to illustrate the various methods discussed and includes appropriate sample spreadsheets. Many are structured so that the reader has the task of finishing them in order to see the results mentioned in the text. It is intended that the readers' curiosity will lead them to explore beyond the given problems.

The purpose of numerical techniques is to provide answers to problems that are either impossible or inconvenient to solve by exact formulae. Nevertheless many of the worked examples and exercises in this book involve problems for which an "exact" solution is known. A first encounter with numerical techniques will be a more positive experience for the learner if he or she can see that the results are in agreement with the exact values.

This book is based on, and goes beyond a second year course taught by the author in the La Trobe University School of Mathematics . The mathematical knowledge assumed is that gained from first-year mathematics, namely basic calculus and elementary matrix algebra.

The author is indebted to those colleagues responsible for the initial assembly of the course in 1988, particularly the author of the original course notes, Dr Peter Forrester, and Professor Edgar Smith and others who have helped teach it. He would like to extend particular gratitude to Professor Robert Smith of Miami University, Ohio, Richard Beare of Warwick University, and Val Dragan, Dr. Arthur Jones, Dr. Alan Andrew, and Dr. Gary Davis at La Trobe for reading parts of the manuscript and subsequent help with errors, typos and matters of style and content. All remaining shortcomings are, of course, to be blamed on the author.

1 Introduction to the Spreadsheet

§1.1   What is a Spreadsheet?

You may have no idea as to the basic structure and function of a typical spreadsheet. If this is the case, it may help if you first consider a simple model for the structure of the spreadsheet, from which it is easy to move to an understanding of a "real" spreadsheet, as implemented on a computer.

Imagine a very large sheet of paper spread out in front of you, on which horizontal and vertical lines have been ruled so as to divide it into many boxes or "cells". Each cell lies in a unique horizontal row and vertical column, and its position in this large table of cells could be described by saying exactly which row and which column. All that is needed is a way of labelling the rows and columns.

A natural thing to do would be to give each row and each column its own number or "index", in a fashion similar to the labelling of rows and columns of matrices. Another method is to label columns with a letter of the alphabet and rows with a number.

If we choose to label the columns with a letter of the alphabet and the rows with a number, then the top left-hand corner of this "spreadsheet" would appear as in Fig 1.1.1.

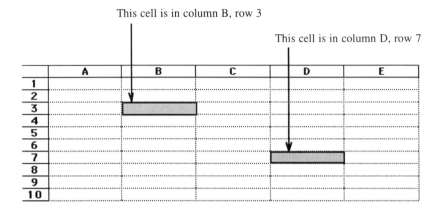

Fig 1.1.1

Each cell has a unique "address" in the spreadsheet. Thus, for example, the two cells indicated in Fig 1.1.1 can be referred to as cell B3 and cell D7, if we adopt the convention that we specify column letter first, then row number.

Imagine further that you can place in any cell either some text, a number, or a formula. In the latter case assume that the data for that formula can be taken from other cells, and that the value produced by a formula can be quickly recalculated should any such data be changed.

Cover this first ruled sheet with another that has a rectangular hole cut in it so that you can only see a few rows and columns of the underlying sheet (through the hole). Move the top sheet parallel to the rows and columns of the lower one, thus choosing which block of rows and columns is visible - as depicted in Fig 1.1.2.

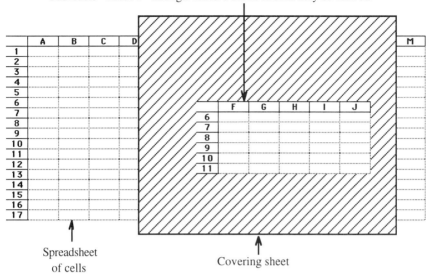

Moveable "window" through which a block of cells may be viewed

Spreadsheet
of cells

Covering sheet

Fig 1.1.2 *A model of the structure of a spreadsheet.*

What you now have is a description of the essential part of the display presented by any (computer software) spreadsheet! For example, the spreadsheet program used to prepare this book, the Macintosh version of Microsoft Excel, appears as shown in Fig. 1.1.3.

From here on, the word **spreadsheet** will be used to refer to such software, rather than to the ruled sheets of paper that served above to introduce the idea behind spreadsheets.

Obviously a computerised version of a sheet of paper with lines ruled on it is not a lot of use without some more special features. What makes a spreadsheet special is what we can put in the cells, and the ease with which quite sophisticated displays can be derived from the stored data.

9

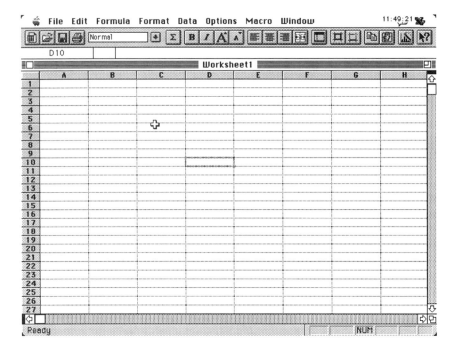

Fig 1.1.3 A typical Microsoft Excel spreadsheet screen display (before entry of any data)

Each cell can store **text** (labels), a **number**, or a **formula**. The data for the formula is obtained from other cells. In other words, what is in one cell might depend on what is in other cells. A change in the value stored in a cell will cause re-calculation in all dependent cells, usually automatically .

The spreadsheet display can usually be set so that either the formulae themselves, or the numerical (or text) values that they produce are visible. Furthermore, the style, font and size of text and the format of numbers can be customized for clarity.

1.1.1 Example

Construct a table of the integer numbers 0 to 10 and their squares.

Fig 1.1.4 shows both the number and formula displays for a spreadsheet that performs this simple task.

	A	B		A	B
1	Table of Squares		1	Table of Squares	
2			2		
3	n	n squared	3	n	n squared
4	0	0	4	0	=A4^2
5	1	1	5	=A4+1	=A5^2
6	2	4	6	=A5+1	=A6^2
7	3	9	7	=A6+1	=A7^2
8	4	16	8	=A7+1	=A8^2
9	5	25	9	=A8+1	=A9^2
10	6	36	10	=A9+1	=A10^2
11	7	49	11	=A10+1	=A11^2
12	8	64	12	=A11+1	=A12^2
13	9	81	13	=A12+1	=A13^2
14	10	100	14	=A13+1	=A14^2

Fig 1.1.4 Two views of a simple spreadsheet, showing numerical values (left) and formulas (right)

Cells A1, A3 and B3 contain text and cell A4 has a number in it. Cells A5 to A14 and B4 to B14 all contain a formula. Note the dependence in these formulae on values stored in other cells. A closer look will quickly reveal the workings of this important aspect of a spreadsheet.

In cell A5, the formula **=A4+1** adds 1 to the value (=0) stored in cell A4 and assigns the result (=1) to cell A5.

In cell A6, the formula **=A5+1** adds 1 to the value (=1) stored in cell A5 and assigns the result (=2) to cell A6, and so on, down to A14.

In cell B4 the formula **=A4^2** squares the value stored in cell A4 and assigns the result to cell B4.

In cell B5 the formula =**A5^2** squares the value stored in cell A5 and assigns the result to cell B5, and so on, down to cell B14.

Another very attractive capability of spreadsheets is the ease with which numerical data can be displayed graphically. Once the data to be displayed is selected in a way appropriate to the particular spreadsheet, a few simple commands will produce the required graph.

Graph 1.1.1 shows a graph of the data in the block of cells A3 to B14 in the spreadsheet of Fig 1.1.4.

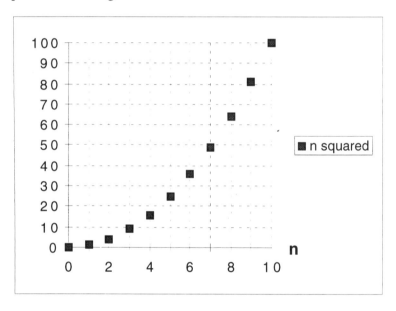

Graph 1.1.1 A graph generated from the spreadsheet of Fig 1.1.4.

The type of entry in a cell is signalled in different ways by different spreadsheets.

For Microsoft® Excel and Claris Resolve™ a formula is preceded by an = sign, and an entry not so preceded will be regarded as text. Numbers need not be preceded by an = sign.

In the case of the IBM PC based spreadsheet Lotus 1-2-3, a formula must begin with a + or a - symbol (unless it begins with a number), and text must be preceded by one of the "label-prefix" characters ' or ^ or ".

The first of these is inserted automatically for any entry that begins with an alphabetic character. This book has been prepared using Microsoft® Excel which can be used on a Macintosh computer or an IBM PC.

Pictures like those in Fig 1.1.4 and Graph 1.1.1 are in fact images captured from a computer screen. In the case of spreadsheet images they will hereafter be referred to as such - for example Screen 1.2.1 below.

Note: Spreadsheets currently available are almost all intended primarily for business use, not as a vehicle for performing mathematics. The default settings for many capabilities such as graph plotting are often unchangeable from a business-oriented type such as bar charts, and the user may wish to customize "template" spreadsheets to get around such difficulties. On the other hand most spreadsheets are very easy to use (once learnt), and little effort is needed to select the desired options in each application.

§1.2   The FILL Command

There is not a lot of typing required to set up the simple spreadsheet depicted in Fig 1.1.4, but this would not be so if instead we had wanted to tabulate the first 100 integers and their squares. A daunting prospect indeed, without the power of the spreadsheet FILL command.

Note: This command is named and effected in various ways according to the particular spreadsheet in use - e.g. with the keystrokes /C when using Lotus 1-2-3, or via ⌘+D or ⌘+R (or mouse and menu to get *Fill Down* or *Fill Right*) when (say) Excel is used on a Macintosh. The word FILL will be used here to indicate this command, to be translated by the reader into the actions appropriate to their particular spreadsheet.

The FILL command is a very convenient way of producing a column or row of very similar formulas, such as those in columns A and B of the spreadsheet of Fig 1.1.4. All that is needed is the first occurrence of the formula in question.

Screen 1.2.1 shows the minimum amount of stuff that has to typed to be able subsequently to create that spreadsheet with the aid of FILL.

	A	B
1	Table of Squares	
2		
3	n	n squared
4	=0	=A4^2
5	=A4+1	
6		

Screen 1.2.1 These formulas (cells A4, A5, B4) must be entered manually.

Now, to get the formulas in cells A6 to A14, first select or "mark" the block of cells A5 to A14. For Excel this is done by dragging the mouse pointer over them. Then execute the **Fill Down** command (⌘+D for Excel on a Macintosh), so that the spreadsheet now appears as in Screen 1.2.2. Excel users can also achieve the same effect by dragging the Fill handle (at the bottom right corner) of cell A5 down to cell A14.

	A	B
1	Table of Squares	
2		
3	n	n squared
4	=0	=A4^2
5	=A4+1	
6	=A5+1	
7	=A6+1	
8	=A7+1	
9	=A8+1	
10	=A9+1	
11	=A10+1	
12	=A11+1	
13	=A12+1	
14	=A13+1	

Screen 1.2.2 Column A after Fill Down has been applied, from A5 to A14.

Similarly, for the formulas in cells B5 to B14, select the block of cells B4 to B14, then use **Fill Down**. The final result is as in Fig 1.1.4. You can see how easy it would be to create a table with many more rows than this.

14

It is possible to FILL more than one column at a time. For this particular example, another way would be first to get the formula =A5^2 into cell B5 - either by just typing it in or by using **Fill Down** on the block B4 to B5 - and then select the block A5 to B14, and apply **Fill Down**. Doing it this way gets both columns of formulas done at once, obviously convenient when many rows are involved.

Note: The **Fill Down** command has automatically incremented the row reference numbers when the formula in cell A5 (or B4) was copied down into the rows below it, thus ensuring that the formulas are doing what we want them to.

There are circumstances where this automatic incrementation is undesireable. For example, first change the specification of the spreadsheet of Fig 1.1.4 so that it tabulates the mth power of the integers 0 to 10, where m is stored as a variable parameter in cell B2 (say). The desired spreadsheet is shown in Screen 1.2.3, with $m = 3$.

	A	B
1	Table of m'th powers	
2	m =	=3
3	n	n to power m
4	=0	=A4^B$2
5	=A4+1	=A5^B$2
6	=A5+1	=A6^B$2
7	=A6+1	=A7^B$2
8	=A7+1	=A8^B$2
9	=A8+1	=A9^B$2
10	=A9+1	=A10^B$2
11	=A10+1	=A11^B$2
12	=A11+1	=A12^B$2
13	=A12+1	=A13^B$2
14	=A13+1	=A14^B$2

Screen 1.2.3 A spreadsheet to tabulate the m'th powers of integers 0 to 10.

The key difference to observe is the cell reference B$2 rather than B2 in the formulas in column B. The effect of the $ before the row reference number 2 is that it prevents the automatic incrementing of the row reference as we **Fill Down** from cell B4 into cells B5 to B14.

Screen 1.2.4 shows what happens if we type the formula =A4^B2 (instead of =A4^B$2) into cell B4 and then **Fill Down**.

	A	B
1	Table of m'th powers	
2	m =	=3
3	n	n to power m
4	=0	=A4^B2
5	=A4+1	=A5^B3
6	=A5+1	=A6^B4
7	=A6+1	=A7^B5
8	=A7+1	=A8^B6
9	=A8+1	=A9^B7
10	=A9+1	=A10^B8
11	=A10+1	=A11^B9
12	=A11+1	=A12^B10
13	=A12+1	=A13^B11
14	=A13+1	=A14^B12

Screen 1.2.4 *The effect of Fill Down (in column B) when B2 is used instead of B$2 in cell B4. Instead of each formula referring to cell B2 for the power as desired, the **Fill Down** has caused B2 to become B3 in the formula in cell B5, B4 in the formula in cell B6, etc.*

There is also a **Fill Right** command which performs the same function for columns as **Fill Down** does for rows. With this command it is the column (letter) reference which is incremented - unless prevented by a preceding $ symbol. Screen 1.2.5 shows what happens when the formulas =A4 and =$A4 are subjected to **Fill Right**.

	C	D	E
4	=A4	=B4	=C4
5	=$A4	=$A4	=$A4

Screen 1.2.5 *The effect of Fill Right on A4 (in cell C4) and $A4 (in C5).*

This is a good point for the reader to practice making tabulations and creating charts from them. A sample is given in Example 1.2.1, and further experience can be gained, based on Exercises 1.2, or on your own ideas.

1.2.1 Example

Tabulate the function $f(x) = x^2 + ax + b$ *with x taking values from -5 to 5, at intervals of 0.1.*

Screen 1.2.6 shows the first and last few lines of the tabulation. There are 101 tabulation points. The values of the two parameters *a* and *b* are stored in cells C1 and C2, respectively. Graph 1.2.1 shows the graph generated from this data for the case *a* =1 and *b* =1.

	A	B	C
1	Quadratic	a =	=1
2	Function	b =	=1
3			
4			
5	x	f(x)=x^2+ax+b	
6	=-5	=A6^2+C$1*A6+C$2	
7	=A6+0.1	=A7^2+C$1*A7+C$2	

	A	B	C
105	=A104+0.1	=A105^2+C$1*A105+C$2	
106	=A105+0.1	=A106^2+C$1*A106+C$2	

	A	B	C
1	Quadratic	a =	1
2	Function	b =	1
3			
4			
5	x	f(x)=x^2+ax+b	
6	-5	21	
7	-4.9	20.11	

	A	B	C
105	4.9	29.91	
106	5	31	

Screen 1.2.6 *Tabulating* $x^2 + ax + b$, *where a and b may be varied.*

17

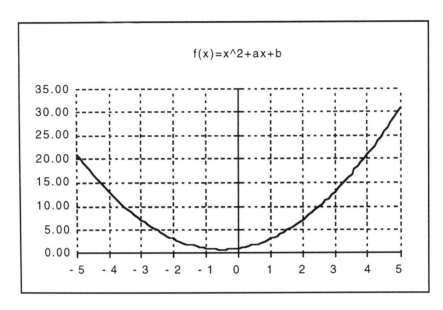

Graph 1.2.1 A graph generated from the spreadsheet of Screen 1.2.6.

Exercises 1.2

(1) Change the spreadsheet of Example 1.2.1 so that the range of x values tabulated begins at a number x_0 that can be varied, with step-size h that may also be varied (instead of the fixed value 0.1). Thus the values of x will be $x_0, x_0 + h, x_0 + 2h, \ldots$. A suitable place to put the parameter values x_0 and h would be the cells C3 and C4, with appropriate labels in cells B3, B4.

(2) Create a spreadsheet that tabulates both the general cubic function $g(x) = ax^3 + bx^2 + cx + d$ and its derivative, $g'(x) = 3ax^2 + 2bx + c$, for the flexible range of x values used in (1) above. Plot the graphs of both $g(x)$ and $g'(x)$ on the same axes (i.e. both in the same chart).

§1.3 Functions Provided with Spreadsheets

The mathematical capabilities of a spreadsheet are numerous. It possesses all the inbuilt functions of a pocket calculator (*sin, cos, log* etc.) so that the formula defined at a cell location can consist of any combination of these functions along with the appropriate use of the arithmetic operations (+, -, *, /, ^) and parentheses. Many spreadsheets also offer a variety of statistical functions and logical functions in addition to business-oriented tools. Some even offer a facility for the user to create their own functions.

A particularly useful function is the **IF** function, which is used in the ordinary everyday sense to choose between two alternate actions, according to some criterion. A simple example will show how it works. First, we will set up a tabulation of a function, and then we will amend that spreadsheet to include a simple application of the IF function.

1.3.1 Example

Tabulate the function $f(x) = x(1-x)$ for $x = -1, 0, 1, 2, 3$.

Screen 1.3.1 shows a suitable spreadsheet.

	A	B			A	B
1	Example 1.3.1			1	Example 1.3.1	
2				2		
3	x	f(x)=x(1-x)		3	x	f(x)=x(1-x)
4	=-1	=A4*(1-A4)		4	-1	-2
5	=A4+1	=A5*(1-A5)		5	0	0
6	=A5+1	=A6*(1-A6)		6	1	0
7	=A6+1	=A7*(1-A7)		7	2	-2
8	=A7+1	=A8*(1-A8)		8	3	-6

Screen 1.3.1 Tabulating $f(x) = x(1-x)$ for $x = -1, 0, 1, 2, 3$.

1.3.2 Example

Suppose now we want to add another column in which it is indicated whether the adjacent value of $f(x)$ is greater than or equal to 0, or not.

This can be effected using the **IF** function, as shown in Screen 1.3.2. An IF function has been used in column C, returning the word "yes" if the adjacent value in column B is non-negative, and "no" otherwise.

	A	B	C
1	Example 1.3.2		
2			
3	x	f(x)=x(1-x)	Is f(x)>=0 ?
4	=-1	=A4*(1-A4)	=IF(B4>=0,"yes","no")
5	=A4+1	=A5*(1-A5)	=IF(B5>=0,"yes","no")
6	=A5+1	=A6*(1-A6)	=IF(B6>=0,"yes","no")
7	=A6+1	=A7*(1-A7)	=IF(B7>=0,"yes","no")
8	=A7+1	=A8*(1-A8)	=IF(B8>=0,"yes","no")

	A	B	C
1	Example 1.3.2		
2			
3	x	f(x)=x(1-x)	Is f(x)>=0 ?
4	-1	-2	no
5	0	0	yes
6	1	0	yes
7	2	-2	no
8	3	-6	no

Screen 1.3.2 *Column C shows use of the **IF** function to choose between two alternatives.*

Consider the formula in cell C4, **=IF(B4>=0, "yes", "no")**. Inspection of the argument of the **IF** function shows it to have **3 parts**:

(1) The first part is a **logical test** which will have the result either *true* or *false*.

(2) The second part **assigns a value** to the cell containing the **IF** when the result of the test is *true*, and

(3) the third part **assigns a value** to the cell containing the **IF** when the result of the test is *false*.

In this example (in cell C4) the logical test is $B4 >= 0$, which asks the question "is the function value stored in cell B4 greater than or equal to 0 ?". The result can only be one or other of the logical values *true* or *false*. In the case at hand, the values assigned are simply the text "yes" when the test is *true*, and the text "no" when it's *false*.

More generally, these two assigned values can be the result of any legitimate spreadsheet calculation, numeric or textual. The general form of the **IF** function, expressed more formally, is

IF(logical test, expression(1), expression(2))

How IF works: should the logical test have the value *true*, then the spreadsheet formula expression(1) is calculated and the cell containing the **IF** function has the resulting numerical (or text) value; conversely, *false* will give the numerical (or text) value computed by expression(2).

Note: Expression(1) and/or expression(2) may also be text valued rather than number valued, as the example above demonstrates.

In other words the **IF** function has the same sense as the everyday use of the word "if" namely: if a certain condition is true do one thing (e.g. evaluate expression(1)) and if it is false do something else (e.g. evaluate expression(2)).

In the example above, expression(1) is the text "yes", and expression(2) is the text "no". These choices could be replaced with any formulas yielding numeric values, for example the simple values 1 and 0 (for the *true*, and *false* cases, respectively).

Note: *Cell references and function names are not case dependent.* Thus, for example, C7 and c7 refer to the same cell, and IF, If, iF and if all specify the IF function.

1.3.3 Exercise

Create a spreadsheet which tabulates the step-function

$$H(x) = \begin{cases} 0 & x < x_0 \\ 1 & x \geq x_0 \end{cases}$$

for $0 \leq x \leq 10$ (say), and $0 < x_0 < 10$ e.g. $x_0 = 5$. Your spreadsheet should cater for easy variation of the value of x_0, similar to the cases of a and b in Example 1.2.1, and for a variable step-size, as in Exercise 1.2(1).

Note: From here on, in most cases, only part of the sample spreadsheet for the examples in this book will be reproduced. It will usually be necessary for the reader to apply the FILL command to their implementation of this minimum of code in order to see the results discussed or displayed in the text.

More Examples using the IF Function

1.3.4 Example

A rectangular pulse function $P(t)$ *can be defined by*

$$P(t) = \begin{cases} 0 & t < T \\ H & T \leq t \leq T + W \\ 0 & t > T + W \end{cases}$$

The pulse, depicted in Fig. 1.3.1, has height H, width W. It "switches on" at $t = T$ *and "switches off" at* $t = T + W$. *Tabulate and display this function, with variable H, W, and T.*

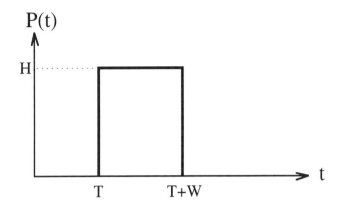

Fig. 1.3.1 *A rectangular pulse function.*

Screen 1.3.3 Shows one way to create this function on a spreadsheet. It arises as the product of two IF functions, the first of which is zero for all $t < T$ and the second for $t > T + W$. Both functions are non-zero (with value 1) only for $T \le t \le T + W$. The values of T, H and W are stored in cells C1, C2 and C3, respectively.

	A	B	C
1	Example 1.3.4	T =	=0.5
2	Rectangular	H =	=1.5
3	Pulse	W =	=1
4		step-size =	=0.05
5			
6	t	P(t)	
7	=0	=IF(A7<C$1,0,1)*IF(A7>C$1+C$3,0,1)*C$2	
8	=A7+C$4	=IF(A8<C$1,0,1)*IF(A8>C$1+C$3,0,1)*C$2	

	A	B	C
1	Example 1.3.4	T =	0.5
2	Rectangular	H =	1.5
3	Pulse	W =	1
4		step-size =	0.05
5			
6	t	P(t)	
7	0	0	
8	0.05	0	

Screen 1.3.3 *Making a rectangular pulse using the product of two IF functions.*

Graph 1.3.1 shows the graph generated from this spreadsheet when $T = 0.5$, $W = 1.0$ and $H = 1.5$. The sides of the pulse are not quite vertical because of the large step-size (as a proportion of T). The reader should try different step-sizes, as well as varying the values of T, W and H.

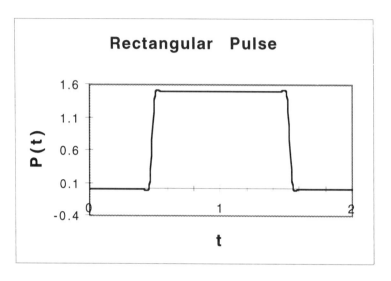

Graph 1.3.1 *A graph of the rectangular pulse from the spreadsheet of Screen 1.3.3.*

There are other ways to achieve the rectangular pulse, using either the logical **OR** function or the logical **AND** function. Instead of using the product of two IF functions,

$$= IF(A7 < C\$1, 0, 1) * IF(A7 > C\$1 + C\$3, 0, 1) * C\$2$$

in cell B7, an equivalent formula is

$$= IF(\ OR(A7 < C\$1, A7 > C\$1 + C\$3), 0, C\$2),$$

and another is

$$= IF(\ AND(A7 >= C\$1, A7 <= C\$1 + C\$3), C\$2, 0).$$

$OR(A7 < C\$1, A7 > C\$1 + C\$3)$ returns the value *true* if either $t < T$ or $t > T + W$, *false* when $T \le t \le T + W$, leading, in turn, to the IF function having value 0 or H, respectively.

$AND(A7 >= C\$1, A7 <= C\$1 + C\$3)$ returns the value *true* if both $t \ge T$ and $t \le T + W$, and *false* otherwise, and the IF function has value H or 0, respectively.

Try these - remember to Fill Down column B after you have changed the formula in cell B7.

Note: The **OR** and **AND** functions are examples of instances where the syntax of the operator varies between spreadsheets. For example with Claris Resolve™ the formula in cell B7 would be

$$= If(A7 < C\$1 \ or \ A7 > C\$1 + C\$3, 0, C\$3)$$

Readers not using Excel should check the syntax of functions and operators in their spreadsheets when transferring formulas from this book.

1.3.5 Example

Tabulate and display the graph of the linear "ramp" function

$$R(t) = \begin{cases} 0 & t < T \\ s(t - T) & t \ge T \end{cases}$$

Screen 1.3.4 shows a suitable spreadsheet, and Graph 1.3.2 shows the graph generated when the slope s of the ramp is 0.5, with $T = 0.5$.

	A	B	C
1	Example 1.3.5	T =	=0.5
2	Ramp Function	s =	=0.5
3		step-size =	=0.05
4			
5	t	R(t)	
6	=0	=IF(A6<C$1,0,C$2*(A6-C$1))	
7	=A6+C$3	=IF(A7<C$1,0,C$2*(A7-C$1))	

	A	B	C
1	Example 1.3.5	T =	0.5
2	Ramp Function	s =	0.5
3		step-size =	0.05
4			
5	t	R(t)	
6	0	0	
7	0.05	0	

Screen 1.3.4 *Spreadsheet for a linear ramp function.*

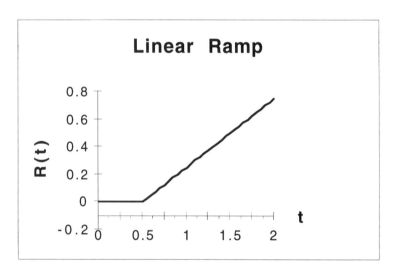

Graph 1.3.2 *A graph of the linear ramp from the spreadsheet of Screen 1.3.4.*

1.3.6 Example

Tabulate and display the "Tent" map $T(x)$, *defined by*

$$T(x) = \begin{cases} 2x & 0 \le x < \frac{1}{2} \\ 2(1-x) & \frac{1}{2} \le x \le 1 \\ 0 & \text{elsewhere} \end{cases}$$

The spreadsheet formula is simplified if this is rewritten as $T(x) = T_1(x) + T_2(x)$, where

$$T_1(x) = \begin{cases} 2x & 0 \le x < \frac{1}{2} \\ 0 & \text{elsewhere} \end{cases}, \quad T_2(x) = \begin{cases} 2(1-x) & \frac{1}{2} \le x \le 1 \\ 0 & \text{elsewhere} \end{cases}$$

For the sake of clarity, the two functions $T_1(x)$ and $T_2(x)$ can be tabulated in separate columns, with the Tent map itself in a third column. The two exclusive intervals $0 \le x < \frac{1}{2}$ and $\frac{1}{2} \le x \le 1$ can be secured using products of IF functions, one of the methods used in Example 1.3.4. This spreadsheet is shown in Screen 1.3.5.

	A	B
1	Tent Map	step-size =
2		
3	x	T1(x)
4	=0	=IF(A4>=0,1,0)*IF(A4<0.5,2*A4,0)
5	=A4+C$1	=IF(A5>=0,1,0)*IF(A5<0.5,2*A5,0)

	C	D
1	=0.02	
2		
3	T2(x)	T(x)
4	=IF(A4>=0.5,1,0)*IF(A4<=1,2*(1-A4),0)	=B4+C4

Screen 1.3.5 *Making the tent function in three stages, using the sum of products of IF functions.*

Of course the three formulas in columns B, C and D can be combined into one in column B. The formula in cell B4 would then be as follows:

$$= IF(A4 >= 0,1,0)*IF(A4 < 0.5,2*A4,0)+$$
$$IF(A4 >= 0.5,1,0)*IF(A4 <= 1,2*(1-A4),0)$$

Alternatively, the logical **AND** function could be used. The formula in cell B4 is then replaced by its equivalent,

$$= IF(AND(A4 >= 0, A4 < 0.5),2*A4,0)+$$
$$IF(AND(A4 >= 0.5, A4 <= 1),2*(1-A4),0)$$

Yet another way is to use the **OR** function, as in Example 1.3.4, in which case the formula in cell B4 would be

$$= IF(OR(A4 < 0, A4 >= 0.5),0,2*A4)+$$
$$IF(OR(A4 < 0.5, A4 > 1),0,2*(1-A4))$$

All of these replacement formulas for cell B4 would, of course, have to be FILLed down into the rows below it.

Graph 1.3.3 is a graph generated from this tabulation.

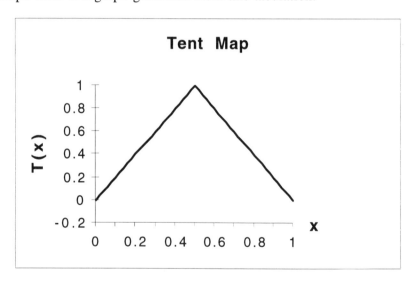

Graph 1.3.3 *A graph of the tent function from the spreadsheet of Screen 1.3.5.*

Exercises 1.3

(1) The linear ramp function of Example 1.3.5 grows for all $t > T$. Modify the spreadsheet so that the function is now defined by

$$R(t) = \begin{cases} 0 & t < T \\ s(t-T) & T \leq t \leq T+W \\ 0 & t > T+W \end{cases}$$

(with $s = 0.5$, $T = 0.5$, $W = 1$ and $0 \leq t \leq 2$, to begin with).

(2) Modify the tent map so that its maximum value (at $t = 0.5$) can be varied.

(3) Create a spreadsheet for the tabulation and graphing of the compound function

$$C(x) = \begin{cases} 0 & x < 1 \text{ or } x > 4 \\ x & 1 \leq x < 2 \\ 2 & 2 \leq x < 3 \\ 2(x-4)^2 & 3 \leq x \leq 4 \end{cases}$$

on the interval $0 \leq x \leq 5$

§1.4 Planning Spreadsheets & Using This Book

1.4.1 Planning

Using a spreadsheet to do numerical mathematics requires a little planning. The sample spreadsheets given with the examples will almost always require the reader to reproduce them on his or her own spreadsheet and then to FILL them as far down as needed to see the results discussed in the text. Obviously this will require that the reader takes the trouble to understand the connection between the mathematics and its "image" in the spreadsheet. Out of this the reader should develop the skills needed to plan and implement extensions to the given spreadsheets and to create original ones, prompted by the exercises and by his or her curiosity.

The use of some low-tech equipment - a ruler, pencil and paper will aid the translation from mathematical symbols to spreadsheet code. The details of how this should be done must lie with the reader.

	A	B	C
1	"Quadratic"	"a ="	1
2	"Function"	"b ="	1
3			
4			
5	"x"	"f(x)=x^2+ax+b"	
6	-5	=A6^2+C$1*A6+C$2	
7	=A6+0.1		

Fig. 1.4.1 Planning a spreadsheet.

An example is offered here for the simple spreadsheet of Screen 1.2.6 above. In Fig. 1.4.1 the entries intended as text (i.e. headings or labels) are shown with inverted commas, while those to become spreadsheet formulae are written as mathematical formulae, at least initially, and then translated to spreadsheet code.

Another way would be to plan the layout of the spreadsheet (as above) with specific details on that plan restricted to the text-bearing cells like A1, A5, etc. above, and to describe elsewhere the details of the mathematical content of the other cells, rows and/or columns, along with the corresponding spreadsheet formulas.

With regard to the example above, the text or "label" cells would be as shown in cells A1, A2, A5, B1, B2 and B5, and the mathematical details would be as follows:

x-values tabulated in column A:
cell A6 formula　　=5 (or just 5),
cell A7 formula　　=A6+0.1 (and FILL down to row 106).

$f(x)$-values tabulated in column B:
cell B6 formula　　=A6^2+C$1*A6+C$2 (and FILL down to row 106).

Value of a in cell C1:　　1

Value of b in cell C2:　　1

Note: Very long formulas are best broken up into smaller parts, trading the use of a few more columns against easier fault tracing.

1.4.2 Why Not Use Sophisticated Facilities?

In this book, spreadsheets are used at their most basic level. Thus, for example, the fact that later releases of Excel allow cells to be given user-defined names has not been exploited here. This facility has the potential for much simplification of the translation of mathematical formulas into spreadsheet formulas, but not all spreadsheets have it.

At several stages in this book a point is reached where further use of a spreadsheet is very difficult or even impossible. In some cases the obstacle may be overcome by use of more sophisticated capabilities (where available), such as user-defined functions and macros, or built-in matrix functions.

These refinements are not pursued here because to do so would defeat the purpose of this book, which is to allow a first exploration of numerical methods without first having to learn any programming techniques. An hour or two is all that is needed to master a spreadsheet to the level needed here.

Of course any reader who does want to go further in their use of spreadsheets for numerical analysis should do so. It is the author's hope that the material in this book will provide him or her with a sound basis from which to begin a deeper exploration.

1.4.3 Using this Book

The material in Chapter 2 (Convergence of Sequences) serves two purposes. Firstly, it provides some more relatively simple exercises with which the unpractised reader can get to know his or her spreadsheet. Secondly, in view of the fact that iterative sequences are a natural part of many of the later topics, it is important that the reader should know that sequences may converge at different rates and that slow ones can often be accelerated.

After Chapter 2 the topics are best taken in the given order, although Chapters 3 and 4 could be exchanged without harm. The sections marked with an asterisk could be regarded as more difficult and be omitted from an introductory course.

It is important for the reader to be aware of certain differences between the way numerical methods are used in a spreadsheet and the way the same methods may be used when implemented using one of the standard numerical computing languages such as FORTRAN or C.

For example, consider the matter or terminating a sequence of calculations, according to a desired accuracy being reached. On a spreadsheet the user sees directly the point at which enough calculations have been done, but in a programmed calculation the program itself must be capable of terminating the calculation. This is all the more important if, as is often the case, the calculation is just a sub-part of a larger computation.

2 Convergence of Sequences

An ordered list of numbers $b_1, b_2, b_3, ..., b_n, ...$ is called a **sequence**, and the typical element b_n is referred to as the nth **term** of the sequence. If the sequence has a finite number of terms it is called a finite sequence. Otherwise it is an infinite sequence, and we will be concerned with the behaviour of some examples of this latter type.

Sequences arise in numerical analysis in many contexts, some of which will be seen in later chapters of this book. In many cases of interest the value of successive terms approaches some particular limiting value which is called the **limit** of the sequence. This property of some infinite sequences is called **convergence**, and the value of the limit is usually the goal of the calculation.

Some convergent sequences converge very slowly, and it may be possible to apply an accelerating transformation in order to reduce the amount of calculation required to find the limit with sufficient accuracy.

This chapter begins with descriptions of some ways in which sequences can be defined and then looks at the concept of convergence, with practical examples for each type of sequence. The speed of convergence of sequences is quantified, and some ways of accelerating slow sequences are considered. The chapter ends with an example that compares some strategies for "stopping" a sequence.

§2.1 Some ways to Define a Sequence

Notation: An infinite sequence having typical term b_n will be written as $\{b_n | n = 1, 2, 3, ...\}$, or simply $\{b_n\}$.

There are many ways in which a sequence may be defined. Three of them, with examples in each case, are:

(i) A simple expression for each term, e.g.

$$b_n = \frac{2n}{n+1}, \quad n = 1, 2, 3, ... \tag{2.1.1}$$

$$a_n = \frac{2(n^2 + 7)}{6n^2 + n + 5}, \quad n = 1, 2, 3, ... \tag{2.1.2}$$

(ii) As a partial sum of a series, e.g.

$$a_n = 1 - \frac{1}{2} + \frac{1}{3} - \frac{1}{4} + ... + \frac{(-1)^{n+1}}{n}, \quad n = 1, 2, 3, ... \tag{2.1.3}$$

$$A_n = 1 + \frac{1}{2^2} + \frac{1}{3^2} + ... + \frac{1}{n^2}, \quad n = 1, 2, 3, ... \tag{2.1.4}$$

(iii) By an iterative scheme, in which a typical term is a function of one or more earlier terms of the sequence, e.g.

$$x_0 = 1, \quad x_{n+1} = \frac{x_n^2 + 3}{2x_n}, \quad n = 0, 1, 2, 3, ... \tag{2.1.5}$$

$$x_0 = 1, \quad x_1 = 2, \quad x_{n+1} = \frac{x_n x_{n-1} + 3}{x_n + x_{n-1}}, \quad n = 1, 2, 3, ... \tag{2.1.6}$$

Sequences of types (ii) and (iii) are of particular prominence in numerical analysis, and will be discussed at greater length later in this chapter. Before that, however we will study the example defined by equation (2.1.1) in order to introduce an important concept, viz. **convergence** of sequences.

2.1.1   Example

Tabulate the sequence defined by equation (2.1.1) for some values of n, along with the magnitude of the difference between b_n and 2.

Screen 2.1.1 shows a spreadsheet that performs the required tabulations.

	A	B	C
1	Example 2.1.1		
2	n	b(n)	|b(n)-2|
3	=1	=2*A3/(A3+1)	=ABS(B3-2)
4	=A3+1	=2*A4/(A4+1)	=ABS(B4-2)
5	=A4+1	=2*A5/(A5+1)	=ABS(B5-2)
6	=A5+1	=2*A6/(A6+1)	=ABS(B6-2)
7	=A6+1	=2*A7/(A7+1)	=ABS(B7-2)
8			
9	=100	=2*A9/(A9+1)	=ABS(B9-2)

	A	B	C
1	Example 2.1.1		
2	n	b(n)	|b(n)-2|
3	1	1.0000	1.0000
4	2	1.3333	0.6667
5	3	1.5000	0.5000
6	4	1.6000	0.4000
7	5	1.6667	0.3333
8			
9	1000	1.9980	0.0020

Screen 2.1.1 *A simple example of a convergent sequence, with limiting value 2.*

Column A contains the values of n (=1, 2, 3,...), beginning with $n = 1$ in cell A3, and column B has the corresponding values of the sequence $b_n = 2n/(n+1)$. Column C lists the magnitude of the difference between b_n and 2.

Apparently, as n becomes large, b_n gets closer and closer to 2. Further, if we want to have $|b_n - 2| < 0.2$ (say), we need only choose any value of $n > 9$, and to achieve $|b_n - 2| < 0.1$ requires $n > 19$. You will have to insert more rows and use **Fill Down** to verify these claims. For large values of n the nth term can most easily be obtained (for this type of sequence) by direct substitution, as in row 9 of the spreadsheet shown here.

In fact for any (small) number $\varepsilon > 0$ we have $|b_n - 2| < \varepsilon$ provided that

$$|b_n - 2| = \frac{2}{n+1} < \varepsilon \quad \Rightarrow \quad n > \frac{2}{\varepsilon} - 1$$

The behaviour shown here is an example of **convergence**: By choosing n sufficiently large we can make b_n as close as we like to 2, and we say that this particular sequence **converges** to the **limit** 2.

§2.2 Convergent Sequences

Intuitively, we say that a sequence $\{b_n\}$ converges to a limit b if, when we take n large enough, b_n becomes as close as we like to b. A more precise statement follows:

2.2.1 Definition: Convergence of a Sequence

Let $\varepsilon > 0$ be given. **The sequence $\{b_n\}$ converges to a limit b** if, for all such ε, there exists an integer $N(\varepsilon)$ such that, for **all** $n > N(\varepsilon)$, $|b_n - b| < \varepsilon$.

Notation: A way of expressing the convergence of a sequence $\{b_n\}$ to limit b is to write

$$\lim_{n \to \infty} b_n = b$$

In words this is: " b_n approaches b as n approaches infinity", or "the limit as n approaches infinity of b_n is b".

We will now explore the convergence of some sequences for which the value of the limit is known. The reader should appreciate that in "real life" applications the limit is *not* known, its value is what is being sought.

2.2.2   Example

Consider the sequence $\{a_n\}$ defined by equation (2.1.2) above:

$$a_n = \frac{2(n^2 + 7)}{6n^2 + n + 5} = \frac{2(1 + \frac{7}{n^2})}{6 + \frac{1}{n} + \frac{5}{n^2}} \to \frac{2}{6} = \frac{1}{3} \text{ as } n \to \infty,$$

i.e. the sequence $\{a_n\}$ has limit $\frac{1}{3}$. Set up a spreadsheet that tabulates both a_n and $\left|a_n - \frac{1}{3}\right|$ against n.

See Screen 2.2.1 where columns A, B, and C respectively list the values of n, a_n and $\left|a_n - \frac{1}{3}\right|$. By inspecting that table we can see, for example, that $\left|a_n - \frac{1}{3}\right| < 0.01$ when $n > 6$, i.e. $N(0.01) = 6$.

Graph 2.2.1 shows plots of the sequence and also of the horizontal lines $\frac{1}{3} + \varepsilon$ and $\frac{1}{3} - \varepsilon$ for the case $\varepsilon = 0.02$, (the value of ε is stored in cell D1 of Screen 2.2.2). You can see that $N(0.02) = 5$. The latter two plots require two extra columns of data, one each for the values $\frac{1}{3} + \varepsilon$ and $\frac{1}{3} - \varepsilon$. These are not included in the spreadsheets shown here - it is left to the reader to include them.

In fact we can build into the spreadsheet a convenient *indicator* of when $\left|a_n - \frac{1}{3}\right| < \varepsilon$, for any ε. For instance, we can set up another column where the formula in each cell gives the text "no" while the preceding inequality is *false* (i.e. $\left|a_n - \frac{1}{3}\right| \geq \varepsilon$), or "yes" when it is *true*, with the value of ε being stored in some convenient cell.

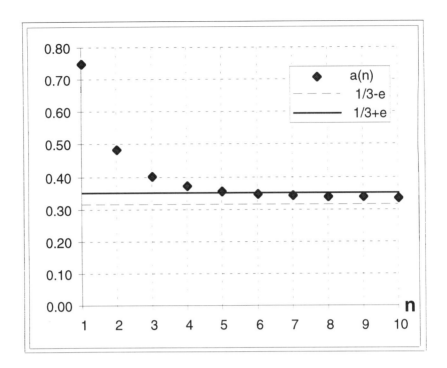

Graph 2.2.1 *A picture of the progress of a convergent sequence to within distance ε from its limit.*

Screen 2.2.2 shows an appropriate modification of Screen 2.2.1 to achieve this end using the spreadsheet **IF** function to test whether $\left| a_n - \tfrac{1}{3} \right| \geq \varepsilon$. The "test" is in cells D6, D7,..., and the quantity $\left| a_n - \tfrac{1}{3} \right|$ is tabulated in cells C6, C7,...

	A	B	C
1	Example 2.2.2		
2		Sequence	
3		a(n)=(2n^2+7)/(n^2+n+5)	
4			
5	n	a(n)	|(a(n)-1/3)|
6	=1	=(2*A6^2+7)/(6*A6^2+A6+5)	=ABS(B6-1/3)
7	=A6+1	=(2*A7^2+7)/(6*A7^2+A7+5)	=ABS(B7-1/3)

Screen 2.2.1 *Tabulating a sequence defined by a simple expression.*

	A	B	C
1	Example 2.2.2		
2		Sequence	
3		a(n)=(2n^2+7)/(n^2+n+5)	
4			
5	n	a(n)	|(a(n)-1/3)|
6	1	0.75000	0.41667
7	2	0.48387	0.15054
8	3	0.40323	0.06989
9	4	0.37143	0.03810
10	5	0.35625	0.02292
11	6	0.34802	0.01468
12	7	0.34314	0.00980
13	8	0.34005	0.00672
14	9	0.33800	0.00467
15	10	0.33659	0.00325

Screen 2.2.1 (continued)

	C	D
1	epsilon =	=0.05
2		
3		
4		
5	|(a(n)-1/3)|	|(a(n)-1/3)|< epsilon?
6	=ABS(B6-1/3)	=IF(C6<D$1,"yes","no")

	A	B	C	D
1	Example 2.2.2		epsilon =	0.05
2		Sequence		
3		a(n) = (2n^2+7)/(n^2+n+5)		
4				
5	n	a(n)	|(a(n)-1/3)|	|(a(n)-1/3)|< epsilon?
6	1	0.75000	0.41667	no
7	2	0.48387	0.15054	no
8	3	0.40323	0.06989	no
9	4	0.37143	0.03810	yes

Screen 2.2.2 *Including an "indicator" in column D, showing when the error is less than ε*

As discussed in Chapter 1, the general form of the **IF** function is

$$\textbf{IF}(\text{logical test, expression}(1), \text{expression}(2))$$

Here, expression(1) and expression(2) are the simple constant (text) values "yes" and "no" respectively, and the logical test gives *true* when the value of $\left| a_n - \frac{1}{3} \right|$ is less than ε. Smaller values of ε may require the FILLing of more rows.

There are further examples of the use of the **IF** function in §1.3.

Exercises 2.2.1

(1) For the sequence tabulated in Screen 2.2.2, find $N(\varepsilon)$ for $\varepsilon = 0.05$, 0.005, 0.001 . You may have to extend your spreadsheet (using FILL) to more rows as you decrease the value of ε. Be sure to save it to a disk file once you're sure it works.

(2) Conduct similar investigations for other simply defined convergent sequences e.g. $(n+3)/(n^2+3)$ (which has limit 0), and $\left(1 + \frac{1}{n}\right)^n$ (which has limit e). One thing that is immediately apparent is that some sequences converge very slowly.

Notes:

(1) An important source of error in computer calculations is the limited precision available for the representation of numbers. A dramatic illustration of this is afforded by further exploration of the last exercise above. To proceed more rapidly along the sequence $\left(1 + \frac{1}{n}\right)^n$, replace n with 10^n and observe what happens as n increases.

(2) Computer-aided explorations like Example 2.2.1 and the Exercises above are intended primarily to encourage an understanding of the concept of convergence. They cannot be used to *prove* that a sequence is convergent. Such proofs usually require the manual application of the appropriate techniques from Analysis.

2.2.3 Sequences defined as Sums:

Suppose we have a sequence $\{a_n\}$. We can define another sequence $\{S_n\}$ by letting the term S_n be the sum of the first n terms of the sequence $\{a_n\}$. Hence

$$S_1 = a_1$$
$$S_2 = a_1 + a_2$$
$$S_3 = a_1 + a_2 + a_3$$
$$\vdots$$
$$S_n = a_1 + a_2 + a_3 + \ldots + a_{n-1} + a_n$$

2.2.4 Definition: Sigma notation for sums:

$$S_n = \sum_{i=1}^{n} a_i = a_1 + a_2 + a_3 + \ldots + a_{n-1} + a_n$$

The sum S_n is called a (finite) **series**, and we will be interested in its behaviour as n is allowed to become very large. Note that

$$S_n = \sum_{i=1}^{n} a_i = (a_1 + a_2 + \ldots + a_{n-1}) + a_n = S_{n-1} + a_n, \quad n \geq 2$$

This gives us an inductive definition of S_n that is convenient for use in a spreadsheet, as you will see in examples below.

2.2.5 Example

A sequence that is defined as a sum of terms was given by equation (2.1.3):

$$a_n = 1 - \frac{1}{2} + \frac{1}{3} - \frac{1}{4} + \ldots + \frac{(-1)^{n+1}}{n} = \sum_{j=1}^{n} \frac{(-1)^{j+1}}{j}$$

which has limit (as $n \to \infty$*) equal to ln2. Tabulate this sequence, which can also be defined inductively by*

$$a_1 = 1, \quad a_n = a_{n-1} + \frac{(-1)^{n+1}}{n}, \quad n \geq 2.$$

The latter form is convenient for a spreadsheet tabulation of this sequence, as shown in Screen 2.2.3. After the initial term $a_1 = 1$ (placed in cell B7), the formula in each of the cells B8, B9,... adds the quantity $(-1)^{n+1} / n$ to the value in the cell immediately above it. Columns A, C and D contain the values of n, $|a_n - \ln 2|$, and the convergence "test", respectively.

In this example the convergence is slow and the successive terms oscillate about the limit.

Exercises 2.2.2

Construct a spreadsheet to examine the convergence of the sequences $\{A_n\}$ and $\{B_n\}$ defined by

(1) $A_n = \sum_{j=1}^{n} \frac{1}{j^2} = A_{n-1} + \frac{1}{n^2}$, which has limit $\frac{\pi^2}{6}$, and

(2) $B_n = \sum_{j=1}^{n} \frac{(-1)^{j+1}}{2j-1}$, which converges to $\frac{\pi}{4}$.

	A	B	C	D
1	Example		eps =	=0.1
2	2.2.5		limit = log2 =	=LN(2)
3		Sequence a(n)=		
4		sum(1 to n)		
5		{(-1)^(j+1)/j}	|e(n)|=	
6	n	a(n)	|a(n)-log2|	|e(n)|< eps?
7	=1	=1	=ABS(B7-D$2)	=IF(C7<D$1,"yes","no")
8	=A7+1	=B7+(-1)^(A8+1)/A8	=ABS(B8-D$2)	=IF(C8<D$1,"yes","no")

	A	B	C	D
1	Example		eps =	0.1
2	2.2.5		limit = log2 =	0.69314718
3		Sequence a(n)=		
4		sum(1 to n)		
5		{(-1)^(j+1)/j}	|e(n)|=	
6	n	a(n)	|a(n)-log2|	|e(n)|< eps?
7	1	1.00000000	0.30685	no
8	2	0.50000000	0.19315	no
9	3	0.83333333	0.14019	no
10	4	0.58333333	0.10981	no
11	5	0.78333333	0.09019	yes

Screen 2.2.3 *A convergent sequence defined by a sum, with limit ln2.*

2.2.6 Sequences defined by Iteration:

Loosely speaking, a sequence $\{x_n\}$ is defined iteratively if the typical term x_i is a function of some of the preceding terms $x_{i-1}, x_{i-2} \ldots$ in the sequence. More concisely, it is defined by an ***n*-point iteration** function if it is given by an expression of the form

$$x_{i+1} = F_i(x_i, x_{i-1}, \ldots x_{i-n+1}), \quad i \geq n$$

In the common case where the iteration function F_i does not change from one iteration to the next i.e. it is independent of i, we have what is called a **stationary** iteration.

We will be mainly concerned with stationary one-point or two-point iterations, i.e. the cases where the iteration formula has the form either

$$x_{i+1} = F(x_i), \quad i \ge 1, \quad \text{or} \quad x_{i+1} = F(x_i, x_{i-1}), \quad i \ge 2.$$

Of course these formulae must be accompanied by initial values (x_1, or x_1 and x_2 respectively) in order to define a unique sequence.

Note: It does not matter what integer is used to label the first element of a sequence. Thus, for example, the sequence $x_{i+1} = F(x_i), \quad i \ge 1$ with $x_1 = a$ is the same as the sequence $x_{i+1} = F(x_i), \quad i \ge 0$ with $x_0 = a$.

2.2.7 Examples

(a) Tabulate the sequence defined by the one-point iteration formula of equation (2.1.5):

$$x_{n+1} = \frac{(x_n)^2 + 3}{2x_n}, \quad x_0 = 1$$

This sequence is convergent to $\sqrt{3}$.

(b) Tabulate the sequence defined by the two-point iteration formula of equation (2.1.6):

$$x_{n+1} = \frac{x_n x_{n-1} + 3}{x_n + x_{n-1}}, \quad x_0 = 1, \quad x_1 = 2$$

also convergent to $\sqrt{3}$.

See Screens 2.2.4 and 2.2.5

	A	B
1	Example 2.2.7a	
2	Sequence x(n+1)=	(x(n)^2 + 3)/(2*x(n))
3	Convergent to sqrt(3) =	=SQRT(3)
4		
5	n	x(n)
6	=0	=1
7	=A6+1	=(B6^2+3)/(2*B6)

	A	B
5	n	x(n)
6	0	1
7	1	2
8	2	1.75

Screen 2.2.4 *A convergent sequence defined by a one-point iteration formula*

	A	B
1	Example 2.2.7b	
2	Sequence x(n+1)=	(x(n)x(n-1) + 3)/(x(n)+x(n-1))
3	Convergent to sqrt(3) =	=SQRT(3)
4		
5	n	x(n)
6	=0	=1
7	=A6+1	=2
8	=A7+1	=(B7*B6+3)/(B7+B6)

	A	B
5	n	x(n)
6	0	1
7	1	2
8	2	1.666666667
9	3	1.727272727

Screen 2.2.5 *A convergent sequence defined by a two-point iteration formula.*

Exercise 2.2.3

Construct a spreadsheet tabulating the sequence defined by

$$x_{n+1} = \sqrt{1-x_n}, \quad x_0 = 0.5$$

which converges to $\frac{1}{2}(-1+\sqrt{5})$.

2.2.8   Divergent Sequences

Many sequences are *not* convergent i.e. there are sequences $\{b_n\}$ for which b_n does *not* come arbitrarily close to a unique finite limit b as $n \to \infty$. To put it another way, we say that the limit does not exist. Such a sequence is called a **divergent** sequence. Divergent behaviour occurs in a variety of ways, some of them illustrated by the following examples.

(a) $a_n = \sqrt{n}$. Obviously $a_n \to \infty$ as $n \to \infty$ i.e. the typical term grows without bound as n increases.

(b) $b_n = (-1)^n$. In this case, b_n is alternately +1 or -1, and so does not approach any single value.

(c) $c_n = (-1)^n n$. Here, successive terms both alternate in sign and have a magnitude that grows without bound.

(d) $d_n = \sum_{m=1}^{n} \frac{1}{m}$.

The terms of this sequence grow without bound, like those of (a) above, only much more slowly. The divergence in this case is easily proven, since

$$d_n > \int_1^{n+1} \frac{1}{x} dx = \ln(n+1) \to \infty \text{ as } n \to \infty$$

In this book we will be dealing with problems where the sequences that we encounter will be convergent, and the occurrence of a divergent sequence will usually indicate that something is wrong.

§2.3   Speed of Convergence

The preceding examples and exercises have made it clear that sequences have widely varying rates of convergence. Later we will discuss and explore techniques for accelerating convergent sequences, so it is important to have a clearly defined measure of the speed of convergence.

2.3.1   Example

To motivate a precise and useful definition of the rate of convergence of a sequence, examine the behaviour of the sequence $\{p_n\}$ defined by $p_n = 10^{-(1+s)^n}$, with $s > 0$. This sequence clearly converges to 0. Tabulate this sequence and experiment with the value of s.

See Screen 2.3.1.

	A	B	C
1	Example 2.3.1	s =	0.5
2			
3	n	p(n)	
4	1	3.16E-02	
5	2	5.62E-03	
6	3	4.22E-04	

	A	B	C
1	Example 2.3.1	s =	=0.5
2			
3	n	p(n)	
4	=1	=10^(-(1+C$1)^A4)	
5	=A4+1	=10^(-(1+C$1)^A5)	
6	=A5+1	=10^(-(1+C$1)^A6)	

Screen 2.3.1 *A sequence with adjustable order of convergence.*

You will see that the convergence can be quite slow (when s is small, e.g. 0.01), or quite fast - as is the case when $s = 0.5$. With values of s that are even larger the convergence is very fast indeed.

In this discussion we mean by "convergence" that the difference between p_n and the limit 0 is less than some arbitrarily chosen small amount, say 10^{-20}.

Now modify the spreadsheet - as shown in Screen 2.3.2 - to include a column that tabulates the quantity $c_n = p_{n+1} / p_n^{\alpha}$, where the parameter α can be varied, taking the value stored in cell C2.

	A	B	C
1	Example 2.3.1	s =	=0.5
2		alpha =	=1.5
3	n	p(n)	c(n)
4	=1	=10^(-(1+C$1)^A4)	=B5/B4^C$2
5	=A4+1	=10^(-(1+C$1)^A5)	=B6/B5^C$2

Screen 2.3.2

With s fixed at some value (say in the range $0 < s \leq 1$), vary the value of α and observe the behaviour of the sequence $\{c_n\}$ in column C.

You should see that for value of α very close to (or equal to) $1+s$, the sequence $\{c_n\}$ appears to be convergent. For $\alpha = 1+s$ it is equal to 1 for all n, an extreme case of convergence!

Now, for each value of s, $\{p_n\}$ is a different sequence with its own speed of convergence, and associated with each sequence $\{p_n\}$ there is another convergent sequence $\{c_n\}$ characterised by a particular value of $\alpha = 1+s$.

Note also that it is the $\{p_n\}$ having larger values of α that converge the most rapidly. The parameter α is a measure of the speed of convergence of the original sequence, and this idea is extended to convergent sequences in general with the following definition:

2.3.2 Definition: Order of convergence

Suppose $\{b_n\}$ is a convergent sequence with limit b and let $e_n = b_n - b$ for each $n \geq 1$. If positive constants λ and α exist such that the sequence $\{c_n\}$ with typical term $c_n = |e_{n+1}| / |e_n|^\alpha$ is convergent, with limit λ,

$$\lim_{n \to \infty} \frac{|e_{n+1}|}{|e_n|^\alpha} = \lambda,$$

then the convergence of $\{b_n\}$ to b is of **order** α with **asymptotic error constant** λ.

Note: e_n is the **error**, being the difference between b_n and the limit b. For the contrived (yet instructive) Example 2.3.1 above we had $p = 0$ and $p_n > 0$ and thus $|e_n| = |p_n - 0| = p_n$.

For $\alpha = 1$ or $\alpha = 2$, the rate of convergence is said to be **linear** or **quadratic**, respectively. Of the two parameters α and λ, the order α has much the stronger effect on the speed of convergence. Let us compare two sequences, one linear and the other quadratic, both having the same values of $|e_0|$ and λ, and ask the question: how large must n be so that $|e_n| < 10^{-8}$ (say)?

We can begin by assuming that $|e_{n+1}| \approx \lambda |e_n|^\alpha$, an approximation that becomes better as n increases. Then

$$|e_1| \approx \lambda |e_0|^\alpha, \quad |e_2| \approx \lambda |e_1|^\alpha \approx \lambda (\lambda |e_0|^\alpha)^\alpha = \lambda^{1+\alpha} |e_0|^{\alpha^2},$$

$$|e_3| \approx \lambda |e_2|^\alpha \approx \lambda \left| \lambda^{1+\alpha} |e_0|^{\alpha^2} \right|^\alpha = \lambda^{1+\alpha+\alpha^2} |e_0|^{\alpha^3}, \text{ etc.},$$

and $|e_n| \approx \lambda^{1+\alpha+\alpha^2+...+\alpha^{n-1}} |e_0|^{\alpha^n} = \lambda^m |e_0|^{\alpha^n}$, with $m = \begin{cases} n & \text{if } \alpha = 1 \\ \dfrac{\alpha^n - 1}{\alpha - 1} & \text{if } \alpha > 1 \end{cases}$,

Suppose now that $|e_0| = 1$ and $\lambda = 0.8$, and consider two examples, with $\alpha = 1$ or $\alpha = 2$:

(i) $\alpha = 1$, $|e_n| \approx (0.8)^n < 10^{-8}$ if $n > \dfrac{-8\log(10)}{\log(0.8)} \approx 82.6$ i.e. $n \geq 83$

(and $\lambda = 0.4$ gives $n \geq 21$)

(ii) $\alpha = 2$, $|e_n| = (0.8)^m$ where $m = \dfrac{2^n - 1}{2 - 1} = 2^n - 1$,

and thus

$$|e_n| = (0.8)^{2^n - 1} < 10^{-8} \text{ if } n > \dfrac{\log\left[\dfrac{\log(0.8 \times 10^{-8})}{\log(0.8)}\right]}{\log(2)} \approx 6.4 \text{ i.e. } n \geq 7$$

(and $\lambda = 0.4$ gives $n \geq 5$)

The quadratically convergent sequence is much the faster of the two. In fact the linear case would need λ to be less than about 0.1 to be as fast as the quadratic, a requirement rarely satisfied in practice. Values of λ around 0.9 are common, in fact.

2.3.3   Example

Perform a numerical exploration to find the order of convergence of the sequence of Example 2.2.5,

$$a_n = \sum_{j=1}^{n} \frac{(-1)^{j+1}}{j}$$

Screen 2.3.3 (cells C3, D3, E6, E7,...) shows the changes to the spreadsheet of Screen 2.2.3 that are needed. The order parameter α is stored in cell D3. Column C lists the values of the quantity called $|e_n|$ above, and column E contains the sequence $c_n = |e_{n+1}| / |e_n|^\alpha$

Try a few values of α to convice yourself that $\alpha = 1$ is appropriate, giving $\lambda \approx 0.99$. You may have to FILL several more rows down each column in order to observe the convergence (or absence of it) in column E.

	C	D	E
1	eps =	=0.1	
2	limit = log2 =	=LN(2)	
3	order a =	=1	
4			
5	$|e(n)|=$		
6	$|a(n)-\log2|$	$|e(n)|<$ eps?	$|e(n+1)|/|e(n)|^a$
7	=ABS(B7-D$2)	=IF(C7<D$1,"yes","no")	=C7/C6^D$3

Screen 2.3.3 *Changes to Screen 2.2.3 to allow estimation of the order of convergence and the asymptotic error constant for a sequence converging to log2.*

Exercises 2.3

(1) Prove that $p_n = 10^{-(1+s)^n}$ has order of convergence $\alpha = 1 + s$.

(2) Extend the spreadsheets of (a) Examples 2.2.7. and (b) Exercise 2.2.3 with a view to finding the rate of convergence of these sequences. Note that the rapid convergence of sequences with higher values of α means that the estimates you will get this way for α and λ will be rather crude.

§2.4   Acceleration of Convergence

Let $\{b_n\}$ be a linearly convergent sequence with limit b. The sequence $\{B_n\}$ defined by

$$B_n = b_n - \frac{\left(b_{n+1} - b_n\right)^2}{b_{n+2} - 2b_{n+1} + b_n} = \frac{b_{n+2}b_n - b_{n+1}^2}{b_{n+2} - 2b_{n+1} + b_n} \qquad (2.4.1)$$

(usually) converges more rapidly to b than does the sequence $\{b_n\}$. By "more rapid" we mean in the sense that

$$\lim_{n \to \infty} \frac{B_n - b}{b_n - b} = 0$$

If we define $\Delta b_n = b_{n+1} - b_n$ (the "forward difference"), and

$$\Delta^2 b_n = \Delta(\Delta b_n) = \Delta b_{n+1} - \Delta b_n = b_{n+2} - 2b_{n+1} + b_n ,$$

we see that equation (2.4.1) is

$$B_n = b_n - \frac{\left(\Delta b_n\right)^2}{\Delta^2 b_n} = A(b_n, b_{n+1}, b_{n+2})$$

This idea is the basis of what is known as **Aitken's Δ^2 method**.

Equation (2.4.1) is obtained by solving the following for b:

$$\frac{b_{n+1} - b}{b_n - b} \approx \lambda \approx \frac{b_{n+2} - b}{b_{n+1} - b}$$

and then setting $B_n = b$.

2.4.1　Example

Apply Aitken's acceleration to the sequence of Example 2.2.5

The addition of another column (see Screen 2.4.1, column F) to the spreadsheet (Screen 2.2.3) of Example 2.2.5 provides a ready demonstration of the acceleration provided by Aitken's method. For example the formula in cell F7 computes the "improved" value B_1 from the original sequence values a_1, a_2, and a_3 that are calculated in cells B7, B8, and B9 respectively. Note that B_n is accurate to 3 decimal places long before a_n is even accurate to 2 places.

	F			F
5	Aitken		5	Aitken
6	B(n)		6	B(n)
7	=B7-(B8-B7)^2/(B9-2*B8+B7)		7	0.7
8	=B8-(B9-B8)^2/(B10-2*B9+B8)		8	0.69047619
9	=B9-(B10-B9)^2/(B11-2*B10+B9)		9	0.694444444
10	=B10-(B11-B10)^2/(B12-2*B11+B10)		10	0.692424242
11	=B11-(B12-B11)^2/(B13-2*B12+B11)		11	0.693589744
12	=B12-(B13-B12)^2/(B14-2*B13+B12)		12	0.692857143
13	=B13-(B14-B13)^2/(B15-2*B14+B13)		13	0.693347339
14	=B14-(B15-B14)^2/(B16-2*B15+B14)		14	0.693003342

Screen 2.4.1 *Acceleration by Aitken's method of the sequence tabulated in column C of Screen 2.2.3.*

Another implementation of this idea is **Steffensen's method** which hinges on the assumption that B_1 is already a better approximation to the limit b than is b_3; thus instead of computing the next terms b_4 (from b_3) and b_5 (from b_4), we use B_1 in place of b_3 to produce the accelerated pair b_4' and b_5'. Hence if the original sequence is defined by the recurrence formula $b_{n+1} = f(b_n)$, the sequence computed is

$$
\begin{array}{lll}
b_1 & b_2 = f(b_1) & b_3 = f(b_2) \\
B_1 = A(b_1, b_2, b_3) & b_4' = f(B_1) & b_5' = f(b_4') \\
B_2 = A(B_1, b_4', b_5') & b_6' = f(B_2) & b_7' = f(b_6')
\end{array}
$$

etc.

2.4.2 Example

Apply both Aitken's and Steffensen's methods to the sequence of Exercise 2.2.3., $x_{n+1} = \sqrt{1 - x_n}$, $x_0 = 0.5$.

See Screen 2.4.2. Columns A and B list n and the unaccelerated sequence. Column C contains the Aitken acceleration, and D, E, and F the Steffensen acceleration terms with the same layout as given above. It is clear that Steffensen's method is the faster of the two.

	A	B	C
1	Example 2.4.2		
2		initial value=	=0.5
3			
4		Iteration	Aitken
5	n	x(n+1)=g(x(n))	acceleration
6	=0	=SQRT(1-C$2)	=B6-(B7-B6)^2/(B8-2*B7+B6)
7	=A6+1	=SQRT(1-B6)	=B7-(B8-B7)^2/(B9-2*B8+B7)

	D	E	F
3		limit =	=(-1+SQRT(5))/2
4	Steffensen's Acceleration		
5	x(0)	x(1)	x(2)
6	=C$2	=SQRT(1-D6)	=SQRT(1-E6)
7	=D6-(E6-D6)^2/(F6-2*E6+D6)	=SQRT(1-D7)	=SQRT(1-E7)
8	=D7-(E7-D7)^2/(F7-2*E7+D7)	=SQRT(1-D8)	=SQRT(1-E8)

	A	B	C	D	E	F
1	Example 2.4.2					
2		initial value=	0.5			
3					limit =	0.61803399
4		Iteration	Aitken	Steffensen's Acceleration		
5	n	x(n+1)=g(x(n))	acceleration	x(0)	x(1)	x(2)
6	0	0.70710678	0.61597960	0.50000000	0.70710678	0.54119610
7	1	0.54119610	0.61671282	0.61498984	0.62049187	0.61604231
8	2	0.67735065	0.61715263	0.61803180	0.61803576	0.61803256

Screen 2.4.2 *Accelerations using both Aitken's & Steffensen's methods, for comparison.*

Exercises 2.4

(1) Modify the spreadsheet of Screen 2.4.2 so as to investigate the rate of convergence of the Aitken's accelerated sequence. Compare with what you found in Exercise 2.3(2)(b).

(2) Repeat this for the first column (D) of terms of Steffensen's acceleration in the spreadsheet of Screen 2.4.2.

Note: Steffensen's method gives quadratic convergence. For a proof of this fact, see Schwarz[1989].

§2.5 The Shanks Transformation

The preceding exercises demonstrate that the Aitken acceleration does not necessarily produce a quadratically convergent sequence from a linear one, while Steffensen's does (not proven here). The nonlinear transformation $S(a_n)$ defined by Aitken's method, viz.,

$$B_n = \frac{a_{n+2}a_n - a_{n+1}^2}{a_{n+2} - 2a_{n+1} + a_n} = S(a_n)$$

is also known as the Shanks transformation.

To gain further acceleration we can apply the transformation again. The main danger in the iteration of an accelerating transformation is the rapid accumulation of rounding errors which have their source in the finite precision of the computer and are compounded by the (ultimately) small differences in the denominator of $S(a_n)$.

2.5.1 Example

Apply the iterated Shanks transformation to the series of Example 2.2.5,

$$a_n = \sum_{j=1}^{n}(-1)^{j+1}/j \quad (\to \ln 2 \text{ as } n \to \infty)$$

See Screen 2.5.1. The formula in cell C5 computes an improved value derived from the original sequence values in cells B5, B6 and B7, as defined by Aitken's method, and so on down column C, exactly as in Example 2.4.1. In column D the same Shanks process is applied again but now the "original" sequence is that listed in column C. Finally, in column E the Shanks process is applied to the sequence in column D.

The original series in column B needs thousands of terms to reach the same accuracy as $S_3(a_5)$ the third Shanks iterate (in column E), which requires only sums up to the first 11 terms to be correct to 7 decimal places.

	A	B	C
1	Example	Iterated Shanks	
2	2.5.1	Transformation	
3		Series for ln(2)=	=LN(2)
4	n	a(n)	S1[a(n)]
5	=1	=1	=(B7*B5-B6^2)/(B7-2*B6+B5)
6	=A5+1	=B5+(-1)^(A6+1)/A6	=(B8*B6-B7^2)/(B8-2*B7+B6)

	D	E
4	S2[a(n)]	S3[a(n)]
5	=(C7*C5-C6^2)/(C7-2*C6+C5)	=(D7*D5-D6^2)/(D7-2*D6+D5)

	A	B	C	D	E
1	Example	Iterated Shanks			
2	2.5.1	Transformation			
3		Series for ln(2)=	0.69314718		
4	n	a(n)	S1[a(n)]	S2[a(n)]	S3[a(n)]
5	1	1.00000000	0.70000000	0.69327731	0.69314887
6	2	0.50000000	0.69047619	0.69310576	0.69314668
7	3	0.83333333	0.69444444	0.69316334	0.69314735

Screen 2.5.1 *Acceleration by means of iterated Shanks transformations.*

Exercises 2.5

Apply the iterated Shanks transformation to the series of (1) Exercise 2.2.2(1) and (2) Exercise 2.2.2(2). What improvement results? Is the order of the accelerated convergence higher than linear?

Note: Following the preceding exercises it should now be apparent that some sequences are resistant to significant acceleration by the methods considered above.

§2.6*   Richardson Extrapolation

Many slowly convergent series that are not significantly accelerated by the Shanks transformation can be treated using Richardson extrapolation.

Let S_n be the nth partial sum, i.e.,

$$S_n = \sum_{i=0}^{n} a_i$$

If it can be shown that for large n the asymptotic form of S_n is

$$S_n \sim Q_0 + Q_1 n^{-1} + Q_2 n^{-2} + \ldots$$

then the expression

$$Q_0 = \sum_{k=0}^{N} \frac{S_{n+k}(n+k)^N (-1)^{k+N}}{k!(N-k)!} \tag{2.6.1}$$

has a value which approaches the limit of the sum as N and n are increased, at a rate that is much faster than the original series.

2.6.1   Example

Apply the Richardson extrapolation to the series

$$S_n = \sum_{j=1}^{n} 1/j^2$$

of Exercise 2.2.2(1). This sequence can also be defined by the recurrence relation $S_1 = 1, \quad S_n = S_{n-1} + 1/n^2$.

From equation (2.6.1) above we get:

For N = 1, $Q_0 = -nS_n + (n+1)S_{n+1}$

For N = 2, $Q_0 = \frac{1}{2}n^2 S_n - (n+1)^2 S_{n+1} + \frac{1}{2}(n+2)^2 S_{n+2}$

For N = 3, $Q_0 = -\frac{1}{6}n^3 S_n + \frac{1}{2}(n+1)^3 S_{n+1} - \frac{1}{2}(n+2)^3 S_{n+2} + \frac{1}{6}(n+3)^3 S_{n+3}$

See Screen 2.6.1. The original sequence is in column B, and cases $N = 1$ and $N = 2$ are in columns C and D.

	A	B	C	D
1	Example	inf sum =		
2	2.6.1	pi^2/6 =	=PI()^2/6	
3		Original	Richardson	
4		Series	Extrapolation	
5	n	S(n)	Case N=1	Case N=2
6	=1	=1	=-A6*B6+A7*B7	=A6^2*B6/2-A7^2*B7+A8^2*B8/2
7	=A6+1	=B6+1/A7^2	=-A7*B7+A8*B8	=A7^2*B7/2-A8^2*B8+A9^2*B9/2

	A	B	C	D	E
1	Example	inf sum =			
2	2.6.1	pi^2/6 =	1.64493407		
3		Original	Richardson		
4		Series	Extrapolation		
5	n	S(n)	Case N=1	Case N=2	Case N=3
6	1	1.00000000	1.50000000	1.62500000	1.64351852
7	2	1.25000000	1.58333333	1.63888889	1.64467593

Screen 2.6.1 *Acceleration of a slowly convergent sequence using Richardson extrapolation.*

Exercises 2.6

(1) Extend the spreadsheet of Screen 2.6.1 to larger values of N.

(2) The sequence

$$b_n = \sum_{m=1}^{n} 1/m$$

does not converge, but the sequence $c_n = b_n - \ln(n)$ is convergent, with limit $\gamma \approx 0.5772$, a number known as Euler's constant. Construct a spreadsheet to observe the divergence of $\{b_n\}$ and modify it to find the value of γ, accurate to 10 decimal places. Use the Shanks transformation or, if necessary, the Richardson extrapolation to accelerate the convergence.

(2) Catalan's constant $G \approx 0.915965$ is the limit of the sequence

$$g_n = \sum_{m=0}^{n} \frac{(-1)^m}{(2m+1)^2}$$

Use the Shanks transformation to find G to 9 decimal places.

(3) In contrast to the sequence of Example 2.2.5, the sequence

$$a_n = \sum_{k=1}^{n} \frac{1}{2^k k}$$

converges fairly quickly to $\ln(2)$, being accurate to 12 decimal places with 37 terms. Use a spreadsheet to tabulate this sequence along with at least the first Shanks transformation. Note that the latter gives *worse* not better accuracy than the original sequence. Hence accelerations must be used with some caution.

(4) The sequence

$$b_n = \sum_{k=1}^{n} \frac{1}{k(4k^2 - 1)}$$

has limit $2\ln(2) - 1$. Apply iterated Shanks acceleration and Richardson extrapolation to this sequence using (say) 20 terms of the original sequence.

(5) Quite simple-looking sequences can exhibit a complex variety of behaviours. An important example in the new field of Chaos Theory is the quadratic map exemplified by the iterative sequence $x_{n+1} = \mu x_n (1 - x_n)$, with $0 < x_0 < 1$. Tabulate this sequence (with the value of μ stored as a parameter) for various values of μ between 0 and 4. Plot a graph of x_n against n for a better idea of what this sequence does.

(6) Tabulate the sequence

$$x_{n+1} = x_n - \frac{x_n(x_n^2 - 1)}{2(2x_n^2 - 1)}, \quad x_0 = \sqrt{\frac{3}{7}}.$$

In another column tabulate the same sequence but with the cells containing the value of x_n rounded to k places rather than x_n itself, with the value of k stored in a suitable cell, and with the initial value also rounded to k places. Set $k = 6, 7, ...12$ in turn and observe the effect on the sequence. The rounding is achieved with the spreadsheet **round** function.

Note: this sequence derives from the Newton method (see Chapter 3) for solving the equation $x^2(x^2 - 1) = 0$, which obviously has solutions $x = 0$ and $x = \pm 1$.

§2.7 Stopping Procedures

In the preceding sections of this chapter, spreadsheet studies of the convergence of various sequences have included a visual aid (the indicator function introduced in Example 2.2.2) to seeing when a specified accuracy was reached. This facility depended on the fact that the limit of the sequence being studied was already known.

When convergent sequences are encountered "in the wild" it is almost always true to say that the limit is *not* known *a priori*. In fact it is the object of the calculation to find the value of the limit. Furthermore, the computations will take place without human intervention, and therefore some form of automatic procedure is needed to stop them when the desired accuracy is reached.

One method is to calculate the change in the sequence at each step and stop the process when this change is less than some preset amount. A second method is to calculate the proportional or percentage change at each step and use this quantity to stop the iteration. The latter method offers more reliable control of precision should the magnitude of the limit vary widely.

2.7.1 Example

Screen 2.7.1 shows a spreadsheet where both these methods can be experimented with.

There are two sequences, $\{a_n\}$ and $\{c_n\}$, tabulated in this spreadsheet: one is the sequence studied first in Example 2.2.5

$$a_n = \sum_{j=1}^{n} \frac{(-1)^{j+1}}{j}$$

which converges slowly to ln2, and the other is defined by $c_n = a_n - d$, where the value of the constant d is stored in cell D3.

Variation of the value of d allows the limit of the second sequence can be set to any value. Column E provides an indicator of whether the difference of successive terms of both sequences is less than ε (stored in cell D1). Columns F and G indicate whether the percentage variation in successive terms is less than the amount P set in cell D2.

	A	B	C	D
1	Example		eps =	=0.05
2	2.7.1		% error P=	=10
3			d=	=0.1
4			D = ln2-d=	=LN(2)-D3
5			b(n) =	c(n) =
6	n	a(n)	a(n)-ln2	a(n)-d
7	=1	=1	=B7-LN(2)	=B7-D$3
8	=A7+1	=B7+(-1)^(A8+1)/A8	=B8-LN(2)	=B8-D$3

	E
5	|c(n)-D|
6	< eps?
7	=IF(ABS(D7-D$4)<D$1,"yes","no")

	F
4	% change
5	in a(n)
6	< P ?
7	
8	=IF(100*ABS((B8-B7)/B8)<D$2,"yes","no")

	G
4	% change
5	in c(n)
6	< P ?
7	
8	=IF(100*ABS((D8-D7)/D8)<D$2,"yes","no")

	A	B	C	D	E	F	G
1	Example		eps =	0.05			
2	2.7.1		% error P=	10			
3			d=	0.1			
4			D = ln2-d=	0.59315		% change	% change
5			b(n)=	c(n)=	|c(n)-D|	in a(n)	in c(n)
6	n	a(n)	a(n)-ln2	a(n)-d	< eps?	< P ?	< P ?
7	1	1.00000	0.30685	0.90000	no		
8	2	0.50000	-0.19315	0.40000	no	no	no

Screen 2.7.1 *A spreadsheet for experimental study of stopping procedures.*

Exercise 2.7

Experiment with various values of d, ε and P. In particular, note the effect on the accuracy of the result of cutting off the sequence with small absolute changes and small proportional changes, especially when the limit of the sequence is itself small.

Perform similar studies using other sequences from earlier examples and exercises.

3 Simultaneous Linear Equations

Simultaneous linear equations occur naturally in many parts of science and engineering. An example is the spectrophotometric analysis of liquid mixtures, considered later in this chapter. The numerical solution of ordinary and partial differential equations often generates a system of linear equations, possibly a very large number of equations, and so the efficient solution of such systems is quite important.

Manual solution of small linear systems can be achieved via the Gaussian Elimination algorithm, a method also available in computer packages for larger systems. It is used here to help illustrate the problem of ill-conditioning of linear systems. This is followed by a consideration of tri-diagonal linear systems, a type which occurs in the numerical solution of some differential equations.

The iterative solution of linear systems is discussed in some detail, including the concept of relaxation techniques and the connection between convergence of an iteration and the eigenvalues of the iteration matrix.

The chapter concludes with some simplified observations on computer arithmetic, with particular reference to ill-conditioned equations and to scaling and pivotting strategies in Gaussian Elimination.

§3.1 Linear Equations and Matrices

We consider here the problem of solving a system of simultaneous linear equations, for example the following three equations with three variables (or "unknowns"):

$$x_1 + x_2 - x_3 = 8$$
$$2x_1 + x_2 + 9x_3 = 12 \qquad (3.1.1)$$
$$x_1 - 7x_2 + 2x_3 = -4$$

More generally, assume that we have N equations with N variables $x_1, x_2, \ldots, x_N$. If the equations are

$$a_{11}x_1 + a_{12}x_2 + \cdots + a_{1N}x_N = b_1$$
$$a_{21}x_1 + a_{22}x_2 + \cdots + a_{2N}x_N = b_2$$
$$\vdots \qquad (3.1.2)$$
$$a_{N1}x_1 + a_{N2}x_2 + \cdots + a_{NN}x_N = b_N$$

then they may be written in matrix form as $\mathbf{Ax} = \mathbf{b}$ where the *coefficient matrix* $\mathbf{A}$ is an $N \times N$ matrix and $\mathbf{x}$ and $\mathbf{b}$ are $N \times 1$ (column) vectors,

$$\mathbf{A} = \begin{bmatrix} a_{11} & a_{12} & \cdots & a_{1N} \\ a_{21} & a_{22} & \cdots & a_{2N} \\ \vdots & \vdots & \ddots & \vdots \\ a_{N1} & a_{N2} & \cdots & a_{NN} \end{bmatrix}, \quad \mathbf{x} = \begin{bmatrix} x_1 \\ x_2 \\ \vdots \\ x_N \end{bmatrix}, \quad \mathbf{b} = \begin{bmatrix} b_1 \\ b_2 \\ \vdots \\ b_N \end{bmatrix}.$$

For the equations (3.1.1) the coefficient matrix $\mathbf{A}$ is given by

$$\mathbf{A} = \begin{bmatrix} 1 & 1 & -1 \\ 2 & 1 & 9 \\ 1 & -7 & 2 \end{bmatrix}, \text{ with } \mathbf{b} = \begin{bmatrix} 8 \\ 12 \\ -4 \end{bmatrix}.$$

Note: We will assume that the system of equations has a unique solution i.e. that the matrix $\mathbf{A}$ is nonsingular, and thus $\det \mathbf{A} \neq 0$.

The *augmented matrix* representing the linear system (3.1.2) is the partitioned matrix

$$
\begin{bmatrix}
a_{11} & a_{12} & \cdots & a_{1N} & b_1 \\
a_{21} & a_{22} & \cdots & a_{2N} & b_2 \\
\vdots & & & & \vdots \\
a_{N1} & a_{N2} & \cdots & a_{NN} & b_N
\end{bmatrix}
$$

If there are only two or three equations it is very little trouble to solve them manually. However as N increases it becomes extremely tedious (to say the least) to do so, and very desirable to have a reliable automated process where the only work required is the input of the coefficients a_{ij} and the b_i terms on the right-hand sides of the equations.

We begin with a direct implementation of a commonly used manual algorithm, Gaussian elimination. The reader will quickly appreciate that this method is very unwieldy to implement on a spreadsheet for any but very small values of N. It is nevertheless included here for several reasons:

(a) It provides a convenient introduction to the problem of ill-conditioned linear systems,

(b) Further appreciation of the deficiencies of computer arithmetic is afforded, and

(c) It affords a useful tool with which to check the solutions provided by other methods that are discussed later on.

Note: For systems of many (perhaps hundreds or more) equations that are represented by a sparse coefficient matrix, the preferred method of solution is by iteration. Sparse matrices have only a small proportion of non-zero elements, for example the matrices that arise in the numerical solution of boundary-value problems and partial differential equations. The use of Gaussian elimination in these cases is wasteful of computer memory and is computationally less efficient, for a given accuracy. Small systems are best solved by Gaussian elimination.

§3.2 Gaussian Elimination

The basic idea behind Gaussian elimination is to apply elementary row operations to the augmented matrix for the system (3.1.2) so as to reduce the left-hand part of it to upper triangular form. The equations can then be solved via back substitution.

An **elementary row operation** is one of the following operations:

(a) the addition of a multiple of any row to another row
(b) the exchange of two rows
(c) multiplication of a row by a (non-zero) constant

A new system of equations created from the old using only elementary row operations will have the same solution set as the original system.

The Gaussian elimination algorithm also provides, as a bonus, the decomposition $A = LU$, where L is a lower triangular matrix and U an upper triangular matrix, with $\det A = \det U = $ product of the diagonal elements of U. Typical forms for L and U are

$$
L = \begin{bmatrix} 1 & 0 & \cdots & 0 \\ l_{21} & 1 & \cdots & 0 \\ \vdots & \vdots & \ddots & \vdots \\ l_{N1} & l_{N2} & \cdots & 1 \end{bmatrix}, \text{ and } U = \begin{bmatrix} u_{11} & u_{12} & \cdots & u_{1N} \\ 0 & u_{22} & \cdots & u_{2N} \\ \vdots & \vdots & \ddots & \vdots \\ 0 & 0 & \cdots & u_{NN} \end{bmatrix}
$$

Once L and U are known the solutions of $Ax = b$ for various b can be found with little extra calculation.

3.2.1　Example

Solve the 3×3 system (3.1.1) manually (using Gaussian elimination), and then set up a spreadsheet for 3×3 systems, using this solved system to test that it works.

The initial augmented matrix is

$$\begin{bmatrix} 1 & 1 & -1 & 8 \\ 2 & 1 & 9 & 12 \\ 1 & -7 & 2 & -4 \end{bmatrix}$$

Stage 1: The aim is to get zeros in the 1st column positions of the 2nd and 3rd rows, with *pivot element* $a_{11} = 1$, and *multipliers* $l_{21} = a_{21} / a_{11} = 2, l_{31} = a_{31} / a_{11} = 1$.

After subtracting $2 \times row(1)$ from row(2), and $1 \times row(1)$ from row(3) the augmented matrix becomes becomes :

$$\begin{bmatrix} 1 & 1 & -1 & 8 \\ 0 & -1 & 11 & -4 \\ 0 & -8 & 3 & -12 \end{bmatrix} \qquad \begin{aligned} R_2' &= R_2 - 2R_1 = R_2 - l_{21}R_1 \\ R_3' &= R_3 - R_1 = R_3 - l_{31}R_1 \end{aligned}$$

(At the side we indicate the required row operations and multipliers l_{ij})

Stage 2: We now need a zero in the 2nd column position of the 3rd row, with pivot element $a_{22}' = -1$, and multiplier $l_{32} = a_{32}' / a_{22}' = 8$.

Subtracting $8 \times row(2)$ from row(3), gives

$$\begin{bmatrix} 1 & 1 & -1 & 8 \\ 0 & -1 & 11 & -4 \\ 0 & 0 & -85 & 20 \end{bmatrix} \qquad R_3' = R_3 - 8R_2 = R_3 - l_{32}R_2$$

The left-hand part of this final augmented matrix is the upper triangular matrix **U**. The multipliers l_{ij} are elements of the lower triangular matrix **L**.

Thus the original equations are equivalent to (i.e. have the same solutions as) the equations

$$x_1 + x_2 - x_3 = 8$$
$$- x_2 + 11x_3 = -4$$
$$- 85x_3 = 20$$

which are easily solved using back substitution (i.e. last equation solved first):

$$x_3 = -20 / 85 = -4 / 17 \approx -0.235294,$$
$$x_2 = 4 + 11x_3 = 24 / 17 \approx 1.411765,$$
$$x_1 = 8 - x_2 + x_3 = 108 / 17 \approx 6.352941$$

The $A = LU$ decomposition for this case has the 3 matrices

$$A = \begin{bmatrix} 1 & 1 & -1 \\ 2 & 1 & 9 \\ 1 & -7 & 2 \end{bmatrix}, \quad L = \begin{bmatrix} 1 & 0 & 0 \\ 2 & 1 & 0 \\ 1 & 8 & 1 \end{bmatrix}, \text{ and } U = \begin{bmatrix} 1 & 1 & -1 \\ 0 & -1 & 11 \\ 0 & 0 & -85 \end{bmatrix}.$$

Exercise: check that $A = LU$.

To see how knowledge of L and U can be used to solve equations that have the same coefficient matrix A but different right-hand sides b, we will first use them to check the solution found above.

With $A = LU$ the equations $Ax = b$ become $Ly = b$, where $Ux = y$. For the case above, $Ly = b$ is

$$\begin{bmatrix} 1 & 0 & 0 \\ 2 & 1 & 0 \\ 1 & 8 & 1 \end{bmatrix} \begin{bmatrix} y_1 \\ y_2 \\ y_3 \end{bmatrix} = \begin{bmatrix} 8 \\ 12 \\ -4 \end{bmatrix},$$

so that

$$y_1 = 8$$
$$2y_1 + y_2 = 12 \ .$$
$$y_1 + 8y_2 + y_3 = -4$$

Solving by forward substitution (i.e. first equation solved first) we get

$$y_1 = 8$$
$$y_2 = 12 - 2y_1 = -4$$
$$y_3 = -4 - y_1 - 8y_2 = 20$$

Next, $\mathbf{Ux} = \mathbf{y}$ is

$$\begin{bmatrix} 1 & 1 & -1 \\ 0 & -1 & 11 \\ 0 & 0 & -85 \end{bmatrix} \begin{bmatrix} x_1 \\ x_2 \\ x_3 \end{bmatrix} = \begin{bmatrix} 8 \\ -4 \\ 20 \end{bmatrix}$$

which we now solve by back substitution as before. To solve another system with the same coefficient matrix $\mathbf{A}$ but different right-hand side $\mathbf{b}$ requires only the forward substitution (for $\mathbf{y}$) then back substitution (for $\mathbf{x}$) steps to be repeated. For example,

$$\mathbf{b} = \begin{bmatrix} 2 \\ 1 \\ 7 \end{bmatrix} \Rightarrow \begin{bmatrix} 1 & 0 & 0 \\ 2 & 1 & 0 \\ 1 & 8 & 1 \end{bmatrix} \begin{bmatrix} y_1 \\ y_2 \\ y_3 \end{bmatrix} = \begin{bmatrix} 2 \\ 1 \\ 7 \end{bmatrix} \Rightarrow \begin{array}{l} y_1 = 2 \\ y_2 = -3 \\ y_3 = 29 \end{array}$$

and then

$$\begin{bmatrix} 1 & 1 & -1 \\ 0 & -1 & 11 \\ 0 & 0 & -85 \end{bmatrix} \begin{bmatrix} x_1 \\ x_2 \\ x_3 \end{bmatrix} = \begin{bmatrix} 2 \\ -3 \\ 29 \end{bmatrix} \Rightarrow \begin{array}{l} x_3 = \frac{-29}{85} \\ x_2 = \frac{-64}{85} \\ x_1 = \frac{41}{17} \end{array}$$

	A	B
1	Example 3.2.1	Gaussian
2		elimination
3	Augmented matrix	
4	=1	=1
5	=2	=1
6	=1	=-7
7	1st stage	
8	=A4	=B4
9	=A5-($A5/$A$4)*A$4	=B5-($A5/$A$4)*B$4
10	=A6-($A6/$A$4)*A$4	=B6-($A6/$A$4)*B$4
11	multipliers l(21)=	=A5/A4
12	2nd stage	
13	=A8	=B8
14	=A9	=B9
15	=A10-($B10/$B9)*A9	=B10-($B10/$B9)*B9
16	multiplier l(32)=	=B10/B9
17	Solutions are	
18	x3=	=D15/C15
19	x2=	=(D14-C14*B18)/B14
20	x1=	=(D13-C13*B18-B13*B19)/A13

	C	D
4	=-1	=8
5	=9	=12
6	=2	=-4
7		
8	=C4	=D4
9	=C5-($A5/$A$4)*C$4	=D5-($A5/$A$4)*D$4
10	=C6-($A6/$A$4)*C$4	=D6-($A6/$A$4)*D$4
11	and l(31)=	=A6/A4
12		
13	=C8	=D8
14	=C9	=D9
15	=C10-($B10/$B9)*C9	=D10-($B10/$B9)*D9
16		
17	det(A) =	=A13*B14*C15

Screen 3.2.1 A spreadsheet that performs Gaussian elimination for a system of 3 equations.

	A	B	C	D
1	Example 3.2.1	Gaussian		
2		elimination	algorithm	
3	Augmented matrix			
4	1	1	-1	8
5	2	1	9	12
6	1	-7	2	-4
7	1st stage			
8	1	1	-1	8
9	0	-1	11	-4
10	0	-8	3	-12
11	multipliers l(21)=	2	and l(31)=	1
12	2nd stage			
13	1	1	-1	8
14	0	-1	11	-4
15	0	0	-85	20
16	multiplier l(32)=	8		
17	Solutions are		det(A) =	85
18	x3=	-0.235294118		
19	x2=	1.411764706		
20	x1=	6.352941176		

Screen 3.2.1 *(continued)*

Screen 3.2.1 shows a spreadsheet set up to solve a 3×3 system using Gaussian elimination. The elements of the augmented matrix are stored in cells A4..D6.

Stage 1 is effected in the block A8..D10, Stage 2 in cells A13..D15, and the back-substitutions for the solution are in cells B18..B20. The determinant det $\mathbf{A}$ = det $\mathbf{U}$ is calculated in cell D17. The numerical values correspond to the test case we have solved above in Example 3.2.1.

Note: We have assumed that the diagonal elements of the coefficient matrix $\mathbf{A}$ are non-zero. If this is not the case then suitable row exchanges are done manually before entry of numbers into the spreadsheet. It is only convenient to incorporate row operations of type (a) into a spreadsheet version of Gaussian elimination. Otherwise, all possible orderings of the rows would have to be catered for.

Exercises 3.2

(1) Set up a spreadsheet like Screen 3.2.1. Use it to solve the system of equations $Ax = b$ defined by

$$A = \begin{bmatrix} 1 & 0 & 2 \\ 3 & 2 & 8 \\ 2 & 3 & 10 \end{bmatrix}, \text{ and } b = \begin{bmatrix} 0 \\ 4 \\ 9 \end{bmatrix},$$

and check by solving manually, first using Gaussian elimination, then (to check it) with the LU decomposition. Next, try the latter method with a different right-hand side b, and see if your manual solution agrees with the results given by the spreadsheet.

(2) Enter the augmented matrix as shown in Screen 3.2.1, but replace the second row with the entries: 2, 1.99, -2.01, 15.8 (instead of 2, 1, 9, 12), and make a note of the solutions, the multipliers and the value of the determinant.

(a) Now change one of the first three column entries in the second row by about 1% (e.g. change -2.01 to -2.03) and note the solutions etc again. What percentage variation in the solutions has been caused by a 1% variation in one of the matrix (A) coefficients?

(b) Compare with the results obtained when the new row(2), column(3) entry is +2.01 (and then varied to +2.03).

Note: There is a simple **geometrical explanation** for the example of the phenomenon seen in Exercise 3.2(2)(a). Each equation is the equation of a plane. Thus to solve the system of 3 linear equations is to find the point(s) of intersection of 3 planes. If two of those planes are almost parallel (as is the case here) then the intersection point will be very sensitive to a change in orientation of either of them. The determinant of the coefficient matrix **A** will be zero if two of the planes are exactly parallel, and small (relative to the average matrix element size) if they are almost parallel. Hence the solution will be less sensitive to errors in the matrix elements if the planes are not nearly parallel. It is sometimes possible to choose them so as to achieve this end, as illustrated in the following example.

§3.3   Ill-conditioned problems

The (first) amended set of equations in Exercise 3.2(2) is an example of what is called an **ill-conditioned** problem: a small variation in a parameter (or an initial condition) defining the system under investigation leads to disproportionately large change in the solution. Ill-conditioning in a linear system is often signalled by a small value for the determinant of the coefficient matrix (relative to the typical element size).

3.3.1 Example

To analyze the composition of a multi-componenent fluid, a possible method is to employ spectrophotometric analysis. This involves the measurement of the absorbance $D(\lambda)$ of light of wavelength λ by the fluid. Beer's law says that for each component the absorbance is proportional to its concentration, with proportionality constant $\mu(\lambda)$. For a three component system the absorbance is given by

$$D(\lambda) = \mu_1(\lambda)c_1 + \mu_2(\lambda)c_2 + \mu_3(\lambda)c_3$$

where μ_1, μ_2, and μ_3 are the proportionality constants for species 1, 2, and 3, and c_1, c_2, c_3 are the respective concentrations. These concentrations can be calculated from the 3×3 system that results from making measurements at three different wavelengths λ_1, λ_2 and λ_3, viz.

$$\begin{bmatrix} \mu_1(\lambda_1) & \mu_2(\lambda_1) & \mu_3(\lambda_1) \\ \mu_1(\lambda_2) & \mu_2(\lambda_2) & \mu_3(\lambda_2) \\ \mu_1(\lambda_3) & \mu_2(\lambda_3) & \mu_3(\lambda_3) \end{bmatrix} \begin{bmatrix} c_1 \\ c_2 \\ c_3 \end{bmatrix} = \begin{bmatrix} D(\lambda_1) \\ D(\lambda_2) \\ D(\lambda_3) \end{bmatrix}$$

In the table below we present data (from Norris [1981]) representing measurements $D(\lambda)$ made at four different wavelengths on a mixture of three components. Find the concentrations, using any 3 rows of data.

λ	μ_1	μ_2	μ_3	$D(\lambda)$
250	0.41	0.47	0.20	1.96
275	0.71	0.29	0.41	2.49
300	0.92	0.40	0.51	3.25
325	0.96	0.19	0.31	2.24

See the following exercise.

Exercise 3.3

Use the data from the first three rows of the table above in the Gaussian elimination spreadsheet of Example 3.2.1. Note the value of the determinant and observe the effect of a small variation of one or more of the coefficient matrix elements. Next, use instead the data from the first, third and fourth rows and repeat the process. You should observe that the former case is ill-conditioned, the latter not so.

•

Note: Some insight can be had here if each row of the augmented matrix is divided by the coefficient entry in that row having the largest magnitude (this is called **scaling**). You will see that the second and third "planes" are almost parallel for the first choice of data.

3.3.2   Round-off (or rounding) Error and Pivoting

An important source of error in the computer implementation of numerical methods is the fact that any given computer has a fixed and finite precision. This means that many numbers will never be represented exactly, and even if two numbers are represented exactly their product (for example) may not be. An important part of the understanding and use of a numerical method for the solution of a problem is the minimization of the potential accumulation of error from this source.

In the example of Gaussian elimination above, the multipliers (e.g. l_{32}) resulted from the division of the leading row elements to be eliminated (e.g. a_{32}) by the so-called **pivot** element (e.g. a_{22}). The first and obvious observation to make is that we cannot permit the pivot entry to be zero. If it is, a suitable row exchange with another row below the pivot row will solve this problem (this is not conveniently accomplished on a spreadsheet).

Next, to help minimise rounding error it is wise to avoid (relatively) small pivots (i.e. large multipliers), again an end that can be achieved by means of a suitable row exchange. This precaution is even more important when the system to be solved is (unavoidably) ill-conditioned. Such "**pivoting strategies**" are an important aspect of the implementation of Gaussian elimination and other similar methods. Detailed discussion and further references may be found in Burden & Faires [1989] and in Schwarz [1989], for instance. As they are not conveniently implemented on a spreadsheet we will not discuss them further here. Some simple examples that emphasize their importance are given at the end of this chapter.

Note: It should be emphasized that these problems are really only significant for "large" systems where the large number of calculations offers greater scope for the accumulation of rounding errors. Later we will discuss iterative (fixed point) methods that are particularly good for large systems and have the attractive property that round-off error can be effectively eliminated.

§3.4   Tri-diagonal Systems

There are many linear systems (i.e. square matrices) that have a particular structure - for example the matrix may be triangular or symmetric . One type that is convenient to solve directly on a spreadsheet is the system having a **tri-diagonal matrix** (defined below), which we will encounter again when we consider solving linear boundary-value problems in chapter 6.

3.4.1 Definition

An $N \times N$ matrix **A** with elements a_{ij} is **tridiagonal** if the only non-zero entries occur for $j = i - 1$ or $j = i$ or $j = i + 1$.

Example

The following 5×5 matrix is tri-diagonal:

$$\begin{bmatrix} 2 & 3 & 0 & 0 & 0 \\ 1 & 2 & -1 & 0 & 0 \\ 0 & -2 & 4 & 1 & 0 \\ 0 & 0 & 3 & 5 & -2 \\ 0 & 0 & 0 & 1 & 1 \end{bmatrix}$$

It is useful to know if a system of interest has a (unique) solution. To this end we define another special sub-class of matrices:

3.4.2 Definition A matrix **A** is **diagonally dominant** if and only if

$$|a_{ii}| \geq \sum_{\substack{j=1 \\ j \neq i}}^{N} |a_{ij}| \quad , \quad i = 1, 2, \dots N$$

A is **strictly diagonally dominant** if the inequality is $>$ instead of $\geq$.

The 5×5 matrix given above is not diagonally dominant.

Example

The matrix below is diagonally dominant but not strictly diagonally dominant:

$$\begin{bmatrix} 2 & 1 & 0 & 1 & 0 \\ 1 & 2 & -1 & 0 & 0 \\ 0 & -2 & 4 & 1 & 0 \\ 0 & 0 & 3 & 5 & -2 \\ 0 & 0 & 0 & 1 & 1 \end{bmatrix}$$

The following theorem guarantees the existence and uniqueness of the solution of a special type of tri-diagonal system:

3.4.3 Theorem A tri-diagonal, diagonally dominant system has a unique solution if the elements a_{ij} of the coefficient matrix $\mathbf{A}$ satisfy the inequalities:

$$a_{ii} < 0, \quad i = 1, 2, \ldots, N$$

and

$$a_{j,j+1} > 0 \quad and \quad a_{j+1,j} > 0, \quad j = 1, 2, \ldots, N-1$$

Note: The theorem above gives *sufficient* conditions for a unique solution - they are not *necessary* conditions. The theorem also holds for the case where all the diagonal elements are positive and the off-diagonal elements negative: simply multiply all the equations by -1 and then use the theorem.

To compute the solution we find the LU decomposition for the coefficient matrix, a relatively simple exercise that does not require Gaussian elimination for a tri-diagonal matrix. If we write

$$\mathbf{A} = \begin{bmatrix} a_{11} & a_{12} & 0 & \cdots & 0 \\ a_{21} & a_{22} & a_{23} & \cdots & 0 \\ \vdots & \vdots & \vdots & \ddots & 0 \\ 0 & 0 & \cdots & a_{N,N-1} & a_{NN} \end{bmatrix} = \mathbf{L}.\mathbf{U}$$

$$= \begin{bmatrix} p_1 & 0 & 0 & \cdots & \cdots \\ a_{21} & p_2 & 0 & \cdots & \cdots \\ 0 & a_{32} & p_3 & 0 & \cdots \\ \vdots & \vdots & \vdots & \ddots & \\ 0 & 0 & \cdots & a_{N,N-1} & p_N \end{bmatrix} . \begin{bmatrix} 1 & q_1 & 0 & 0 & \cdots \\ 0 & 1 & q_2 & 0 & \cdots \\ \vdots & & \ddots & & \\ 0 & \cdots & & 1 & q_{N-1} \\ 0 & \cdots & & 0 & 1 \end{bmatrix}$$

then it follows that, when $\mathbf{A}$ is tri-diagonal,

$$p_1 = a_{11} \qquad\qquad q_1 = \frac{a_{12}}{p_1}$$

$$p_2 = a_{22} - a_{21}q_1$$

$$\vdots \qquad\qquad\qquad q_2 = \frac{a_{23}}{p_2}$$

$$\vdots$$

$$\vdots \qquad\qquad\qquad \vdots$$

$$\vdots \qquad\qquad\qquad q_{N-1} = \frac{a_{N-1N}}{p_{N-1}}$$

$$p_N = a_{NN} - a_{NN-1}q_{N-1}$$

Exercise: Verify this result.

The system $\mathbf{Ax} = \mathbf{b}$ is now $\mathbf{Lz} = \mathbf{b}$ where $\mathbf{z} = \mathbf{Ux}$. The equations $\mathbf{Lz} = \mathbf{b}$ are solved for $\mathbf{z}$ by forward substitution

$$z_1 = \frac{b_1}{p_1}$$

$$z_2 = \frac{(b_2 - a_{21}z_1)}{p_2}$$

$$\vdots$$

$$z_N = \frac{(b_N - a_{NN-1}z_{N-1})}{p_N}$$

and then $\mathbf{Ux} = \mathbf{z}$ is solved for $\mathbf{x}$ using backward substitution:

$$x_N = z_N$$

$$x_{N-1} = z_{N-1} - x_N q_{N-1}$$

$$\vdots$$

$$x_2 = z_2 - x_3 q_2$$

$$x_1 = z_1 - x_2 q_1$$

The algorithm described above for solving a tri-diagonal system is known as Crout's method.

3.4.4 Example

Use Crout's method to solve the 6×6 tri-diagonal system of equations represented by

$$
A = \begin{bmatrix} 3 & 1 & 0 & 0 & 0 & 0 \\ 1 & 3 & -1 & 0 & 0 & 0 \\ 0 & -1 & 3 & 1 & 0 & 0 \\ 0 & 0 & 1 & 3 & 1 & 0 \\ 0 & 0 & 0 & 1 & 3 & 1 \\ 0 & 0 & 0 & 0 & 1 & 3 \end{bmatrix}, \quad b = \begin{bmatrix} 2 \\ -3 \\ 3 \\ -1 \\ 1 \\ -2 \end{bmatrix}
$$

See Screen 3.4.1 and the following explanation of the spreadsheet layout. The solution is $x_1 = x_3 = x_5 = 1, x_2 = x_4 = x_6 = -1$.

The fact that the tridiagonal matrix A is easily mapped to 3 columns (one each for the sub-diagonal, diagonal and super-diagonal entries) has been exploited to ease the work of extending the spreadsheeet to solve a larger system. The table below shows how this has been done for this example

	A	B	C	D	E	F	G	H
5		a_{11}	a_{12}	b_1	p_1	q_1	z_1	x_1
6	a_{21}	a_{22}	a_{23}	b_2	p_2	q_2	z_2	x_2
7	a_{32}	a_{33}	a_{34}	b_3	p_3	q_3	z_3	x_3
8	a_{43}	a_{44}	a_{45}	b_4	p_4	q_4	z_4	x_4
9	a_{54}	a_{55}	a_{56}	b_5	p_5	q_5	z_5	x_5
10	a_{65}	a_{66}		b_6	p_6		z_6	x_6

	A	B	C	D
1	Example	3.4.4		
2	Matrix	elements		
3			super-	RHS
4	sub-	diags	diags	of eqns
5	diags	=3	=1	=2
6	=1	=3	=-1	=-3
7	=-1	=3	=1	=3
8	=1	=3	=1	=-1
9	=1	=3	=1	=1
10	=1	=3		=-2

	E	F	G	H
4	p(i)	q(i)	z(i)	x(i)
5	=B5	=C5/E5	=D5/E5	=G5-H6*F5
6	=B6-A6*F5	=C6/E6	=(D6-A6*G5)/E6	=G6-H7*F6
7	=B7-A7*F6	=C7/E7	=(D7-A7*G6)/E7	=G7-H8*F7
8	=B8-A8*F7	=C8/E8	=(D8-A8*G7)/E8	=G8-H9*F8
9	=B9-A9*F8	=C9/E9	=(D9-A9*G8)/E9	=G9-H10*F9
10	=B10-A10*F9		=(D10-A10*G9)/E10	=G10

	A	B	C	D	E	F	G	H
1	Example	3.4.4						
2	Matrix	elements			Crout's	Method		
3			super-	RHS				
4	sub-	diags	diags	of eqns	p(i)	q(i)	z(i)	x(i)
5	diags	3	1	2	3.0000	0.3333	0.6667	1
6	1	3	-1	-3	2.6667	-0.3750	-1.3750	-1
7	-1	3	1	3	2.6250	0.3810	0.6190	1
8	1	3	1	-1	2.6190	0.3818	-0.6182	-1
9	1	3	1	1	2.6182	0.3819	0.6181	1
10	1	3		-2	2.6181		-1.0000	-1

Screen 3.4.1 *Solving a tri-diagonal system of 6 linear equations using Crout's method.*

Exercise 3.4

Adapt the spreadsheet of Screen 3.2 to solve the 5×5 system:

$$\mathbf{A} = \begin{bmatrix} -2 & 1 & 0 & 0 & 0 \\ 1 & -2 & 1 & 0 & 0 \\ 0 & 1 & -2 & 1 & 0 \\ 0 & 0 & 1 & -2 & 1 \\ 0 & 0 & 0 & 1 & -2 \end{bmatrix} \quad \mathbf{b} = \begin{bmatrix} 1 \\ 0 \\ 0 \\ 0 \\ 0 \end{bmatrix}$$

§3.5 Solving Linear Systems by Iteration

To solve a problem by **iteration** means to repeatedly apply an appropriate rule (e.g. a recurrence formula) until the difference between the results of successive applications is sufficiently small.

In the case of the linear system $\mathbf{Ax = b}$ starting with an approximate solution $\mathbf{x}^{(0)}$, we may generate a sequence of vectors $\mathbf{x}^{(0)}, \mathbf{x}^{(1)}, \mathbf{x}^{(2)}, \mathbf{x}^{(3)}, \ldots$, convergent to the solution $\mathbf{x}$. The existence and uniqueness of, and convergence to the solution can sometimes be established from the imposition of conditions on the matrix elements, similar to those used above in the case of tri-diagonal systems.

3.5.1 The Jacobi and Gauss-Seidel methods:

For any matrix with non-zero diagonal elements (e.g. a diagonally dominant matrix) the system of equations (3.1.2) can be re-written as

$$x_1 = \frac{1}{a_{11}}(b_1 - a_{12}x_2 - a_{13}x_3 - \cdots - a_{1N}x_N)$$

$$x_2 = \frac{1}{a_{22}}(b_2 - a_{21}x_1 - a_{23}x_3 - \cdots - a_{2N}x_N)$$

$$\vdots$$

$$x_N = \frac{1}{a_{NN}}(b_N - a_{N1}x_1 - a_{N2}x_2 - \cdots - a_{NN-1}x_{N-1})$$

which can be summarised as

$$x_i = \frac{1}{a_{ii}}(b_i - \sum_{\substack{j=1 \\ j \neq i}}^{N} a_{ij}x_j) \quad , \quad 1 \leq i \leq N$$

Jacobi's method turns these equations into an iterative scheme as follows:

$$x_i^{(n+1)} = \frac{1}{a_{ii}}(b_i - \sum_{\substack{j=1 \\ j \neq i}}^{N} a_{ij}x_j^{(n)}), \quad 1 \leq i \leq N \qquad (3.5.1)$$

Here, $x_i^{(n)}$ is the i'th component of the vector $\mathbf{x}^{(n)}$ in the sequence $\mathbf{x}^{(0)}, \mathbf{x}^{(1)}, \ldots, \mathbf{x}^{(n)}, \ldots$

When there are three equations we have

$$a_{11}x_1 + a_{12}x_2 + a_{13}x_3 = b_1 \Rightarrow x_1^{(n+1)} = \frac{1}{a_{11}}(b_1 - a_{12}x_2^{(n)} - a_{13}x_3^{(n)})$$

$$a_{21}x_1 + a_{22}x_2 + a_{23}x_3 = b_2 \Rightarrow x_2^{(n+1)} = \frac{1}{a_{22}}(b_2 - a_{21}x_1^{(n)} - a_{23}x_3^{(n)}) \quad (3.5.2)$$

$$a_{31}x_1 + a_{32}x_2 + a_{33}x_3 = b_3 \Rightarrow x_3^{(n+1)} = \frac{1}{a_{33}}(b_3 - a_{31}x_1^{(n)} - a_{32}x_2^{(n)})$$

For example, for the equations

$$2x_1 + x_2 + 2x_3 = 7$$
$$4x_1 + 5x_2 + x_3 = 8$$
$$x_1 + 2x_2 + 6x_3 = 9$$

the Jacobi iteration equations are

$$x_1^{(n+1)} = (7 - x_2^{(n)} - 2x_3^{(n)})/2$$
$$x_2^{(n+1)} = (8 - 4x_1^{(n)} - x_3^{(n)})/5$$
$$x_3^{(n+1)} = (9 - x_1^{(n)} - 2x_2^{(n)})/6$$

3.5.2 Example

Apply Jacobi's method to the system of three equations:

$$2x_1 + x_2 + 2x_3 = 7$$
$$4x_1 + 5x_2 + x_3 = 8$$
$$x_1 + 2x_2 + 6x_3 = 9$$

Screen 3.5.1 shows a spreadsheet that uses these formulae to solve the given equations, using the starting values (1.2,0.8,0.9), stored in cells A8..C8. Formulas corresponding to equations 3.5.2 are in cells A9..C9, and these are FILLed Down the spreadsheet to produce the iteration. The exact solution is $(\frac{33}{13}, \frac{-9}{13}, \frac{17}{13})$ which is, to 4 decimal places, (2.5385, -0.6923, 1.3077). The formulae include frequent use of the $ symbol to help reduce the amount of typing needed, aided by editing and the FILL command.

Screen 3.5.1 also shows another iterative scheme, the **Gauss-Seidel** method. Here, once the next iterate of a vector component $x_i^{(n+1)}$ has been computed it is then used in the calculation of all subsequent vector components.

The equation for $x_1^{(n+1)}$ is the same for both methods. For a system of three equations, the Gauss-Seidel expression for $x_2^{(n+1)}$ is

$$x_2^{(n+1)} = \frac{1}{a_{22}}(b_2 - a_{21}x_1^{(n+1)} - a_{23}x_3^{(n)})$$

which differs from the Jacobi formula (given earlier) at the second term on the right-hand side, and

$$x_3^{(n+1)} = \frac{1}{a_{33}}(b_3 - a_{31}x_1^{(n+1)} - a_{32}x_2^{(n+1)})$$

which differs in both the second and third terms on the right-hand side.

	A	B
1	Example 3.5.2	
2	Augmented matrix	
3	=2	=1
4	=4	=5
5	=1	=2
6		Jacobi
7	x(1)	x(2)
8	=1.2	=0.8
9	=($D$3-$B$3*B8-$C$3*C8)/$A$3	=($D$4-$A$4*A8-$C$4*C8)/$B$4

	C	D
3	=2	=7
4	=1	=8
5	=6	=9
6		
7	x(3)	n
8	=0.9	=0
9	=($D$5-$A$5*A8-$B$5*B8)/$C$5	=D8+1

	E
7	x(1)
8	=A8
9	=($D$3-$B$3*F8-$C$3*G8)/$A$3

	F
7	x(2)
8	=B8
9	=($D$4-$A$4*E9-$C$4*G8)/$B$4

	G
7	x(3)
8	=C8
9	=($D$5-$A$5*E9-$B$5*F9)/$C$5

	A	B	C	D	E	F	G
1	Example 3.5.2						
2	Augmented matrix						
3	2	1	2	7			
4	4	5	1	8			
5	1	2	6	9		Gauss-	
6		Jacobi				Seidel	
7	x(1)	x(2)	x(3)	n	x(1)	x(2)	x(3)
8	1.20000	0.80000	0.90000	0	1.20000	0.80000	0.90000
9	2.20000	0.46000	1.03333	1	2.20000	-0.34000	1.24667

Screen 3.5.1 Jacobi and Gauss-Seidel iterations for a system of 3 equations.

The Gauss-Seidel iteration formulas are in cells E9..G9 (for the first step) and the block below (for further iteration), with the starting values copied from those used for the Jacobi iteration.

In the case of this particular example it is clear that the Gauss-Seidel iteration settles down more quickly than the Jacobi iteration. In general this is the case, but *not always*.

A measure of how "good" an approximate solution x^* is for the linear system $Ax = b$ is provided by the **residual vector r** defined by

$$r = b - Ax^*$$

The aim, then, of the iterative method is to produce a sequence of residual vectors $r^{(0)}, r^{(1)}, r^{(2)}, \ldots$ that converges quickly to zero. Here, $r^{(n)} = b - Ax^{(n)}$.

For a given matrix A and initial vector $x^{(0)}$, there may be theoretical prescriptions for the nature of A that imply convergence. One result is

---

3.5.3 Theorem

If A is strictly diagonally dominant, then for any choice of $x^{(0)}$ both the Jacobi and Gauss-Seidel methods give sequences $x^{(0)}, x^{(1)}, x^{(2)}, x^{(3)}, \ldots$ that converge to the unique solution of $Ax = b$.

In Example 3.5.2 the matrix A is not even diagonally dominant, let alone strictly so. Experimentation with different initial vectors in this case (try it!) often shows wild transient behaviour in the Jacobi sequence and quick convergence in the Gauss-Seidel.

Another theoretical result, one that allows a comparison of the two methods, for some systems, is:

3.5.4 Theorem (Stein -Rosenberg)

If $a_{ij} \leq 0$ for $i \neq j$ and $a_{ii} > 0$ for $i = 1, 2, 3, \ldots, N$ then the Gauss-Seidel and Jacobi iterations either both converge or both diverge, and in the former case the Gauss-Seidel method converges faster.

Exercises 3.5

(1) Apply both iterations to systems with the coefficient matrices listed below, including a computation of the residual vector. Choose your own right-hand side entries **b** for the equations, and get the "exact" solutions from your Gaussian elimination spreadsheet. Note which (if any) method gives convergence to the known solution.

$$\begin{bmatrix} 1 & -2 & 2 \\ -1 & 1 & -1 \\ -2 & -2 & 1 \end{bmatrix}, \quad \begin{bmatrix} 1 & -2 & 1 \\ -1 & 1 & -1 \\ -2 & -2 & 1 \end{bmatrix}, \quad \begin{bmatrix} 1 & \frac{1}{2} & \frac{1}{2} \\ -1 & 1 & -1 \\ -\frac{1}{2} & \frac{1}{2} & 1 \end{bmatrix}$$

(2) Use iterative methods to solve the 6×6 system of Example 3.4.4 and the 5×5 system of Exercise 3.4. Work directly with the particular equations, rather than including the entire augmented matrix (as in the preceding examples).

Note: For the purpose of introducing iterative methods it has been convenient to study systems of just a few equations. In practice, as mentioned earlier, these methods are particularly well suited to the solution of large, sparse systems of equations.

§3.6* Relaxation methods

To gain some insight into how the iterative methods discussed above work, it will prove useful to express the coefficient matrix **A** as the sum of three matrices - **L**, **D** and -**R** as follows:

$$\mathbf{A} = -\mathbf{L} + \mathbf{D} - \mathbf{R}, \text{ where } \begin{cases} d_{ij} = a_{ij}, \ i = j & and \quad d_{ij} = 0, \ i \neq j \\ l_{ij} = -a_{ij}, \ i > j & and \quad l_{ij} = 0, \ i \leq j \\ r_{ij} = -a_{ij}, \ i < j & and \quad r_{ij} = 0, \ i \geq j \end{cases}$$

For the 3×3 case these are

$$D = \begin{bmatrix} a_{11} & 0 & 0 \\ 0 & a_{22} & 0 \\ 0 & 0 & a_{33} \end{bmatrix}, L = \begin{bmatrix} 0 & 0 & 0 \\ -a_{21} & 0 & 0 \\ -a_{31} & -a_{32} & 0 \end{bmatrix}, R = \begin{bmatrix} 0 & -a_{12} & -a_{13} \\ 0 & 0 & -a_{23} \\ 0 & 0 & 0 \end{bmatrix}.$$

Example

In the case given in example 3.5.2 we have

$$D = \begin{bmatrix} 2 & 0 & 0 \\ 0 & 5 & 0 \\ 0 & 0 & 6 \end{bmatrix}, \quad L = \begin{bmatrix} 0 & 0 & 0 \\ -4 & 0 & 0 \\ -1 & -2 & 0 \end{bmatrix}, \quad R = \begin{bmatrix} 0 & -1 & -2 \\ 0 & 0 & -1 \\ 0 & 0 & 0 \end{bmatrix}.$$

The decomposition of A defined above allows the **Jacobi** iteration formulae of equation (3.5.1) to be written as

$$\begin{aligned} x^{(n+1)} &= D^{-1}b + D^{-1}(L+R)x^{(n)} \quad (= C_J x^{(n)} + D^{-1}b, \text{ say}) \\ &= x^{(n)} + D^{-1}(L-D+R)x^{(n)} + D^{-1}b \\ &= x^{(n)} + D^{-1}r^{(n)} \end{aligned}$$

Exercise *

Verify this claim. You may find it convenient to study a 3×3 system first.

●

This last formula says that the next iterate is obtained from the last by adding the "correction" $D^{-1}r^{(n)}$. It is reasonable to suppose that the speed of convergence might be affected by varying the *amount* of this correction that is added. Thus a natural modification is to introduce a **relaxation parameter** ω as follows:

$$\mathbf{x}^{(n+1)} = \mathbf{x}^{(n)} + \omega \mathbf{D}^{-1} \mathbf{r}^{(n)}$$

$$= \mathbf{C}_J(\omega)\mathbf{x}^{(n)} + \omega \mathbf{D}^{-1}\mathbf{b}$$

where

$$\mathbf{C}_J(\omega) = (1 - \omega)\mathbf{I} + \omega \mathbf{D}^{-1}(\mathbf{L} + \mathbf{R})$$

Hence

$$x_i^{(n+1)} = (1 - \omega)x_i^{(n)} + \omega(b_i - \sum_{\substack{j=1 \\ j \neq i}}^{N} a_{ij}x_j^{(n)})/a_{ii} \quad , \quad 1 \leq i \leq N \quad (3.6.1)$$

This is known as the **simultaneous relaxation (SR) method.** When $\omega = 1$ we have the original Jacobi method, and it is possible that faster convergence may occur for $\omega \neq 1$. For $\omega < 1$ the method is referred to as simultaneous under-relaxation, while for $\omega > 1$ it is called simultaneous over-relaxation (or SOR).

An important question that arises immediately is this: what value of ω maximises the rate of convergence? This is not a simple problem. At this point we will experiment with SR on a spreadsheet for a 3×3 system, for which the SR formulae are

$$x_1^{(n+1)} = (1 - \omega)x_1^{(n)} + \frac{\omega}{a_{11}}(b_1 - a_{12}x_2^{(n)} - a_{13}x_3^{(n)})$$

$$x_2^{(n+1)} = (1 - \omega)x_2^{(n)} + \frac{\omega}{a_{22}}(b_2 - a_{21}x_1^{(n)} - a_{23}x_3^{(n)}) \qquad (3.6.2)$$

$$x_3^{(n+1)} = (1 - \omega)x_3^{(n)} + \frac{\omega}{a_{33}}(b_3 - a_{31}x_1^{(n)} - a_{32}x_2^{(n)})$$

3.6.1 Example

Use the Jacobi iteration with relaxation to solve the equations in Example 3.5.2. Begin with same initial point as used in that example, and $\omega = 1$.

Screen 3.6.1 shows a spreadsheet with the relaxation method, including columns E to G for the residual and its magnitude (column H) at each step. The value of the relaxation parameter ω is stored in cell C2, the starting values in cells A10..C10, and the iteration formulas corresponding to equations 3.6.2 begin in cells A11..C11 and continue in the cells below these three.

	A
9	x(1)
10	=1.2
11	=(1-C$2)*A10+C$2*($D$4-$B$4*B10-$C$4*C10)/$A$4

	B
9	x(2)
10	=0.8
11	=(1-C$2)*B10+C$2*($D$5-$A$5*A10-$C$5*C10)/$B$5

	C	D
9	x(3)	n
10	=0.9	=0
11	=(1-C$2)*C10+C$2*($D$6-$A$6*A10-$B$6*B10)/$C$6	=D10+1

	E	F
9	r(1)	r(2)
10	=D$4-(A$4*A10+B$4*B10+C$4*C10)	=D$5-(A$5*A10+B$5*B10+C$5*C10)

	G	H
9	r(3)	magnitude
10	=D$6-(A$6*A10+B$6*B10+C$6*C10)	=SQRT(E10^2+F10^2+G10^2)

Screen 3.6.1 Jacobi iteration for 3 equations, with variable relaxation parameter ω in cell C2.

	A	B	C	D
1	Example 3.6.1	Simultaneous relaxation		
2		parameter $\omega =$ 0.83		
3	Augmented matrix			
4	2	1	2	7
5	4	5	1	8
6	1	2	6	9
7				
8		Jacobi with relaxation		
9	x(1)	x(2)	x(3)	n
10	1.20000	0.80000	0.90000	0
11	2.03000	0.51780	1.01067	1
12	2.19636	-0.09966	0.99274	2

	D	E	F	G	H
8			residual		
9	n	r(1)	r(2)	r(3)	magnitude
10	0	2.00000	-1.70000	0.80000	2.74408
11	1	0.40087	-3.71967	-0.12960	3.74345
12	2	0.72147	-1.27985	1.04654	1.80382

Screen 3.6.1 (*continued*)

Exercise 3.6.1

A simple experiment is to observe the size of the residual as ω is varied, starting at $\omega = 1$, and going up or down in steps of (say) 0.1 initially, and then with smaller steps. It seems that the best convergence obtains for $\omega = 0.83$, approximately.

The **Gauss-Seidel** iteration formula can be expressed as

$$x_i^{(n+1)} = \frac{1}{a_{ii}}(b_i - \sum_{\substack{j=1 \\ j<i}}^{n} a_{ij}x_j^{(n+1)} - \sum_{\substack{j=1 \\ j>i}}^{n} a_{ij}x_j^{(n)})$$

from which it follows that

$$\mathbf{x}^{(n+1)} = (\mathbf{D}-\mathbf{L})^{-1}\mathbf{R}\mathbf{x}^{(n)} + (\mathbf{D}-\mathbf{L})^{-1}\mathbf{b}$$
$$= \mathbf{C}_G\mathbf{x}^{(n)} + (\mathbf{D}-\mathbf{L})^{-1}\mathbf{b}$$

Simultaneous relaxation can be introduced here as follows:

$$\mathbf{x}^{(n+1)} = \mathbf{C}_G(\omega)\mathbf{x}^{(n)} + \omega(\mathbf{D}-\omega\mathbf{L})^{-1}\mathbf{b}$$

where

$$\mathbf{C}_G(\omega) = \omega(\mathbf{D}-\omega\mathbf{L})^{-1}[(1-\omega)\mathbf{D}+\omega\mathbf{R}].$$

For simplicity we will discuss the 3×3 case. We use

$$x_1^{(n+1)} = (1-\omega)x_1^{(n)} + \frac{\omega}{a_{11}}(b_1 - a_{12}x_2^{(n)} - a_{13}x_3^{(n)})$$

$$x_2^{(n+1)} = (1-\omega)x_2^{(n)} + \frac{\omega}{a_{22}}(b_2 - a_{21}x_1^{(n+1)} - a_{23}x_3^{(n)}) \qquad (3.6.3)$$

$$x_3^{(n+1)} = (1-\omega)x_3^{(n)} + \frac{\omega}{a_{33}}(b_3 - a_{31}x_1^{(n+1)} - a_{32}x_2^{(n+1)})$$

Note: This is the original Gauss-Seidel iteration when $\omega = 1$. As expected, the only difference from the Jacobi iteration (with relaxation) is that on the right-hand side we use those components of $\mathbf{x}^{(n+1)}$ that have already been computed.

3.6.2 Example

Repeat the previous example but using the formulae for Gauss-Seidel iteration (with relaxation).

See Screen 3.6.2. Only the first three columns (containing the iteration formulas corresponding to equations 3.6.3) are shown, since the rest of the cells are the same as in Screen 3.6.1 .

	A
9	x(1)
10	=1.2
11	=(1-C$2)*A10+C$2*($D$4-$B$4*B10-$C$4*C10)/$A$4
12	=(1-C$2)*A11+C$2*($D$4-$B$4*B11-$C$4*C11)/$A$4

	B
8	Gauss-Seidel with relaxation
9	x(2)
10	=0.8
11	=(1-C$2)*B10+C$2*($D$5-$A$5*A11-$C$5*C10)/$B$5
12	=(1-C$2)*B11+C$2*($D$5-$A$5*A12-$C$5*C11)/$B$5

	C
9	x(3)
10	=0.9
11	=(1-C$2)*C10+C$2*($D$6-$A$6*A11-$B$6*B11)/$C$6
12	=(1-C$2)*C11+C$2*($D$6-$A$6*A12-$B$6*B12)/$C$6

	A	B	C
8	Gauss-Seidel with relaxation		
9	x(1)	x(2)	x(3)
10	1.200000	0.800000	0.900000
11	2.200000	-0.340000	1.246667
12	2.423333	-0.588000	1.292111

Screen 3.6.2 *Gauss-Seidel iteration with relaxation (showing changes to Screen 3.6.1 only).*

Exercises 3.6.2

(1) Experiment with different values of the relaxation parameter (in Example 3.6.2). There is a theorem (due to W.Kahan) which says that the condition $0 < \omega < 2$ is necessary for the convergence of the (Gauss-Seidel) relaxation method. In the case to hand. it is interesting to note that when $\omega = 1$ (i.e. ordinary Gauss-Seidel) the 3rd component rapidly becomes exact (try it), but the "best" overall rate of convergence is obtained with $\omega \approx 1.1$.

(2) Try the relevant relaxation method in those cases which did not converge in Exercises 3.5.

§3.7* The Iteration Matrix & Rates of Convergence.

Observe that the iteration formulae above all have the matrix form

$$\mathbf{x}^{(n+1)} = \mathbf{C}\mathbf{x}^{(n)} + \mathbf{c}$$

The matrix $\mathbf{C}$ is called the **iteration matrix**. The convergence of a particular method for a given coefficient matrix $\mathbf{A}$ requires that the **spectral radius** $\rho(\mathbf{C})$ of the iteration matrix should satisfy $\rho(\mathbf{C}) < 1$. The spectral radius of a matrix is the modulus of its largest eigenvalue. Thus the problem of whether or not a given method applied to a given set of equations will provide convergence becomes the question (for appropriate matrix $\mathbf{C}$) "is $\rho(\mathbf{C}) < 1$?" If a relaxation method is contemplated we have the problem of minimization of $\rho(\mathbf{C})$ against ω, too difficult to study here.

To get some feel for the effect of the spectral radius we will confine ourselves to a system of 3 equations and consider the ordinary Jacobi and Gauss-Seidel iterations, with no relaxation parameter.

For a system of 3 equations the **Jacobi** iteration matrix is

$$\mathbf{C}_J = \begin{bmatrix} 0 & a & b \\ c & 0 & d \\ e & f & 0 \end{bmatrix}$$

where

$$a = \frac{-a_{12}}{a_{11}}, \quad b = \frac{-a_{13}}{a_{11}}, \quad c = \frac{-a_{21}}{a_{22}}, \quad d = \frac{-a_{23}}{a_{22}}, \quad e = \frac{-a_{31}}{a_{33}}, \quad f = \frac{-a_{32}}{a_{33}}$$

The characteristic equation $\det(\mathbf{C}_J - \lambda\mathbf{I}) = 0$ for this matrix is

$$f(\lambda) = \lambda^3 - (ac + be + df)\lambda - (ade + bcf) = 0 \qquad (3.7.1)$$

In the case of **Gauss-Seidel** iteration we have the matrix

$$\mathbf{C}_G = (\mathbf{D} - \mathbf{L})^{-1}\mathbf{R} = \begin{bmatrix} a_{11} & 0 & 0 \\ a_{21} & a_{22} & 0 \\ a_{31} & a_{32} & a_{33} \end{bmatrix}^{-1} \begin{bmatrix} 0 & -a_{12} & -a_{13} \\ 0 & 0 & -a_{23} \\ 0 & 0 & 0 \end{bmatrix} = \begin{bmatrix} 0 & a & b \\ 0 & c & d \\ 0 & e & f \end{bmatrix}$$

where

$$a = \frac{-a_{12}}{a_{11}}, \quad b = \frac{-a_{13}}{a_{11}}, \quad c = \frac{a_{12}a_{21}}{a_{11}a_{22}}, \quad d = \frac{a_{13}a_{21} - a_{23}a_{11}}{a_{11}a_{22}},$$

$$e = \frac{a_{12}(a_{22}a_{31} - a_{21}a_{32})}{a_{11}a_{22}a_{33}}, \quad \text{and } f = \frac{a_{13}(a_{22}a_{31} - a_{21}a_{32}) + a_{11}a_{23}a_{32}}{a_{11}a_{22}a_{33}}$$

In this case the characteristic equation is

$$\lambda[\lambda^2 - (c+f)\lambda + cf - de] = \lambda g(\lambda) = 0$$

$$\Rightarrow \lambda = 0 \text{ or } g(\lambda) = \lambda^2 - (c+f)\lambda + cf - de = 0 \qquad (3.7.2)$$

It follows that the **spectral radius** will be the magnitude of the root with largest magnitude of the cubic equation (3.7.1) for a Jacobi iteration, or of the quadratic equation (3.7.2) for a Gauss-Seidel iteration.

Equation (3.7.1) is a cubic of the form $f(x) = x^3 + px + q = 0$ and we will first assume that $p \neq 0$ and $q \neq 0$. Let

$$\Delta = -4p^3 - 27q^2, \text{ with } r = -\tfrac{1}{2}q + \tfrac{1}{6}\left(\tfrac{-\Delta}{3}\right)^{\frac{1}{2}}, \text{ and } s = -\tfrac{1}{2}q - \tfrac{1}{6}\left(\tfrac{-\Delta}{3}\right)^{\frac{1}{2}}.$$

There are 3 cases to consider:

(a) $\Delta > 0$ (3 distinct real roots. N.B. $\Delta > 0 \Rightarrow p < 0$)

Here, r and s are the conjugate pair $-\tfrac{1}{2}q \pm \tfrac{1}{6}\left(\tfrac{\Delta}{3}\right)^{\frac{1}{2}}i$.

If we let $Rcis(\theta) = -\tfrac{1}{2}q + \tfrac{1}{6}\left(\tfrac{\Delta}{3}\right)^{\frac{1}{2}}i, \quad 0 < \theta < \pi$, then $R = \sqrt{-p^3/27}$

and if $t = \tan(\theta) = -\dfrac{\left(-\tfrac{\Delta}{3}\right)^{\frac{1}{2}}}{3q}$, then $\theta = \begin{cases} \tan^{-1}(t) & \text{if } q < 0 \\ \pi + \tan^{-1}(t) & \text{if } q > 0 \end{cases}$.

The 3 real roots are then

$$x_1 = 2R^{\frac{1}{3}}\cos(\tfrac{\theta}{3}), \quad x_2 = 2R^{\frac{1}{3}}\cos(\tfrac{\theta+2\pi}{3}), \quad x_3 = 2R^{\frac{1}{3}}\cos(\tfrac{\theta+4\pi}{3}).$$

(b) $\Delta < 0$ (1 real, 2 complex roots) Both r and s are real, and the roots are

$$x_1 = r^{\frac{1}{3}} + s^{\frac{1}{3}}, \quad x_2 = -\tfrac{1}{2}(r^{\frac{1}{3}} + s^{\frac{1}{3}}) + i\tfrac{\sqrt{3}}{2}(r^{\frac{1}{3}} - s^{\frac{1}{3}}), \quad x_3 = \bar{x}_2$$

and the root x_1 has the largest magnitude of the three.

(c) $\Delta = 0$ (3 real roots, at least 2 equal) $r = s = -\tfrac{1}{2}q$ and the roots are

$$x_1 = 2r^{\frac{1}{3}}, \quad x_2 = x_3 = -r^{\frac{1}{3}}.$$

According to which of these 3 cases apply, the root with greatest magnitude ρ is given by:

$$\rho = \begin{cases} Max\left[2R^{\frac{1}{3}}\left|\cos(\tfrac{\theta}{3})\right|, 2R^{\frac{1}{3}}\left|\cos(\tfrac{\theta+2\pi}{3})\right|, 2R^{\frac{1}{3}}\left|\cos(\tfrac{\theta+4\pi}{3})\right|\right] & \text{if } \Delta > 0 \\ \left|r^{\frac{1}{3}} + s^{\frac{1}{3}}\right| & \text{if } \Delta < 0 \\ 2\left|r^{\frac{1}{3}}\right| & \text{if } \Delta = 0 \end{cases}$$

	A	B
1	Cubic x^3+px+q=0	
2		
3		
4	delta > 0	
5	R =	=IF(D3>0,SQRT(-D1^3/27),0)
6	t =	=IF(D3>0,-SQRT(D3/3)/(3*D2),0)
7	theta =	=IF(D2<0,ATAN(B6),PI()+ATAN(B6))
8	theta/3 =	=B7/3
9	cos term(1) =	=ABS(COS(B8))
10	cos term(2) =	=ABS(COS((B7+2*PI())/3))
11	cos term(3) =	=ABS(COS((B7+4*PI())/3))
12	spec rad =	=2*B5^(1/3)*MAX(B9,B10,B11)
13	FINAL spec rad =	=IF(D3>0,B12,D10)

	C	D
1	p =	=-19/30
2	q =	=17/60
3	delta =	=-4*D1^3-27*D2^2
4	delta <= 0	
5	r =	=IF(D3>0,0,-D2/2+SQRT(-D3/3)/6)
6	s =	=IF(D3>0,0,-D2/2-SQRT(-D3/3)/6)
7	r^(1/3) =	=IF(D5>0,D5^(1/3),-(-D5)^(1/3))
8	s^(1/3) =	=IF(D6>0,D6^(1/3),-(-D6)^(1/3))
9		
10	spec rad =	=ABS(D7+D8)

	A	B	C	D
1	Cubic x^3+px+q=0		p =	-0.633333333
2			q =	0.283333333
3			delta =	-1.151351852
4	delta > 0		delta <= 0	
5	R =	0	r =	-0.038416169
6	t =	0	s =	-0.244917164
7	theta =	3.141592654	r^(1/3) =	-0.337420411
8	theta/3 =	1.047197551	s^(1/3) =	-0.625661946
9	cos term(1) =	0.5		
10	cos term(2) =	1	spec rad =	0.963082357
11	cos term(3) =	0.5		
12	spec rad =	0		
13	FINAL spec rad =	0.963082357		

Screen 3.7.1 *Spectral radius of the iteration matrix for Jacobi iteration of 3 equations.*

These results for the Jacobi iteration are implemented in the spreadsheet of Screen 3.7.1. The IF functions in cells B5..B7 (for $\Delta > 0$) and in cells

D5..D8 (for $\Delta \leq 0$) select the various cases discussed, and the IF in cell B13 selects the appropriate result for the spectral radius. The cases where either $p = 0$ or $q = 0$ are easily solved without resorting to a computer:

If $p = 0$, then $\rho = \left| q^{\frac{1}{3}} \right|$ or, if $q = 0$, then $\rho = |p|^{\frac{1}{2}}$. These two special cases can easily be added to the spreadsheet (an Exercise for the reader).

Exercise

Modify the spreadsheet of Screen 3.7.1 to allow direct entry of the elements of the coefficient matrix, thus avoiding the need for manual calculation of the coefficients $p = -(ac + be + df)$ and $q = -(ade + bcf)$ in the cubic equation (3.7.1). Extend the spreadsheet to include a calculation of the spectral radius of the Gauss-Seidel iteration for the same coefficient matrix, derived from equation (3.7.2) and the work preceding it.

3.7.1 Example

Apply the theory and the spreadsheet discussed above, to the coefficient matrix in Example 3.5.2.

For the coefficient matrix of Example 3.5.2 the cubic equation relevant to the Jacobi iteration matrix is $f(\lambda) = \lambda^3 - \frac{19}{30}\lambda + \frac{17}{60} = 0$, and the spreadsheet gives the spectral radius to be $\rho_J \approx 0.9631$. The quadratic equation in the case of the Gauss-Seidel iteration is $g(\lambda) = \lambda^2 - \frac{11}{30}\lambda + \frac{1}{60} = 0$. This equation has roots $\lambda = \frac{1}{60}(11 \pm \sqrt{61})$ and so the spectral radius is $\rho_G = \frac{1}{60}(11 + \sqrt{61}) \approx 0.3135$. These results cast some light on the behaviour of the two methods observed earlier: $\rho_G < \rho_J < 1$ and so both are convergent, with Gauss-Seidel being faster. The fact that $\rho_J \approx 1$ explains the observed sensitivity to the starting point and transient behaviour of the Jacobi iteration.

Exercises 3.7

Find the spectral radii for the coefficient matrices of Exercises 3.5. Are they consistent with the observed behaviour of the relevant iterations?

§3.8 Some Comments on Scaling and Pivoting.

Floating Point Representation of Numbers

In a computer numbers are ultimately represented in binary form, being expressed in terms of the base number 2. The principles of the following discussion are not altered by a change of number base, so we will use the more convenient decimal base number 10.

Non-integer numbers are represented in **floating point** form. For example the number 256.2 is represented as the pair of numbers 0.2562 (called the **mantissa**) and +3 (called the **exponent**), the relationship being given by $number = mantissa \times 10^{exponent}$.

The exponent is chosen so that $0.1 \le |mantissa| < 1.0$ In this example we have $256.2 = 0.2562 \times 10^{+3}$, which we can write as 0.2562 E+03. This way of representing numbers allows a very wide range of number magnitudes, controlled by the exponent, with the precision being determined by that of the mantissa. More examples are:

$37.86 = 0.3786$ E+02, $0.008729 = 0.8729$ E -02, $-9784.1 = -0.97841$ E+04

The **precision** of the mantissa - meaning the number of digits possible - is generally a fixed quantity for a given computer or software, and it is from this restriction that much of the numerical error we encounter arises. The first thing to be considered is to consider rules by which a computer may remove excess digits. Two possibilities are **truncation** by **rounding** or by **chopping**.

Suppose that we wish to represent the number 57.6392. With no precision restriction at all on the mantissa this would be 0.576392 E+02. However, with only (say) 4-digit precision available we can use either

(a) **chopping**: ignore all digits after the 4th, giving 0.5763 E+02

(b) **rounding**: round up the 4th digit if the 5th digit is greater than or equal to 5, which gives 0.5764 E+02.

In neither case do we have the original number, and a loss of precision has occurred. Of course all modern computers and software are equiped with much higher floating point precision than 4 digits, yet they must suffer from this defect as its existence is not dependent on the *degree* of precision available. We will continue to illustrate related computer error difficulties with low precision examples for the sake of convenience.

Floating Point Arithmetic

For **addition** and **subtraction**, numbers are first given the same exponent, namely that of the one having the exponent with largest (absolute) value. This may lead to further loss of precision.

Example

Perform the operation 2750 - 87.1 using 3 digit floating point arithmetic with chopping. The exact difference is 2662.9.

Now 2750 = 0.275 E+04, and 87.1 = 0.871 E+02 ≈ 0.008 E+04, the latter step being the necesssary precursor to the subtraction:

0.275 E+04 - 0.008 E +04 = 0.267 E +04 = 2670 i.e. an error of 7.1.

With rounding instead of chopping we have 87.1 ≈ 0.009 E+04 and the final result then is 0.266 E+04 = 2660.

Even though both original numbers could be represented exactly in the prescribed arithmetic, their difference could not be found correctly.

Multiplication and **division** can be performed without prior adjustment of the exponent of either operand. Usually, for example, in exact arithmetic the product of two numbers with 3 digit precision will result in a number with up to 6 significant digits and error is inevitable if the representation in use is of lower precision.

Example

Calculate the product $4120 \times 83 = 341960$ using 3 digit floating point arithmetic with chopping.

Now, $4120 = 0.412$ E+04 and $83 = 0.830$ E+02, and thus

$$4120 \times 83 = 0.412 \times 0.83 \times 10^6 \approx 0.342 \times 10^6 = 342000$$

Catastrophic Cancellation

Significant errors can be introduced when floating point arithmetic is used to subtract numbers that are nearly equal, especially when amplified by subsequent calculations such as the division of another number by the result.

Example

Find $\sqrt{10020} - 100.1$ using 4 digit arithmetic with rounding.

$$\sqrt{10020} - 100.1 = 100.09995... - 100.1$$
$$\approx 0.1001E + 02 - 0.1001E + 02$$
$$= 0.0000E + 02$$

A better idea of the potential problems caused by this type of error is afforded by a simple example using the Gaussian elimination algorithm. Consider the pair of equations

$$\begin{bmatrix} 0.002 & 1 \\ 1 & 1 \end{bmatrix} \begin{bmatrix} x_1 \\ x_2 \end{bmatrix} = \begin{bmatrix} 1 \\ 2 \end{bmatrix}.$$

Using *exact* arithmetic we have

$$\left[\begin{array}{cc|c} 0.002 & 1 & 1 \\ 1 & 1 & 2 \end{array}\right]_{R_2' = R_2 - \frac{1}{0.002} R_1} \rightarrow \left[\begin{array}{cc|c} 0.002 & 1 & 1 \\ 0 & -499 & -498 \end{array}\right].$$

Solving by back-substitution,

$$x_2 = \frac{498}{499} \text{ and } x_1 = \frac{500}{499}.$$

Now we will do the same calculation but using 2 digit floating point arithmetic with chopping: the computation of the entry a_{22}' in row 2, column 2 given by the row operation

$$R_2' = R_2 - \frac{1}{0.002} R_1$$

is now

$$\begin{aligned} a_{22}' &= 1 - \tfrac{1}{.002} \times 1 \\ &= 0.001 \times 10^3 - 0.5 \times 10^3 \\ &\approx 0.00 \times 10^3 - 0.50 \times 10^3 \quad \text{(2 digit arithmetic)} \\ &= -500. \end{aligned}$$

Next, back-substitution gives

$$x_2 = \frac{0.49 \times 10^3}{0.50 \times 10^3} = 0.98$$

which is in good agreement with the exact answer, and

$$x_1 = \frac{(1 - x_2)}{0.002} = \frac{1}{0.002}(0.10 \times 10^1 - 0.09 \times 10^1) = \frac{(0.01 \times 10^1)}{0.20 \times 10^{-2}} = 50,$$

which is way off the correct value.

To trace the error to its source, reconsider the step

$$a'_{22} = a_{22} - \frac{a_{21}}{a_{11}} \times a_{12} = 1 - \frac{1}{0.002} \times 1$$

$$= 0.001 \times 10^3 - 0.5 \times 10^3$$

$$= 0.00 \times 10^3 - 0.50 \times 10^3$$

$$= -500.$$

The error is the loss of the term $0.001 \times 10^3 (= a_{22})$, apparently because the relatively large size of the multiplier $a_{21} / a_{11} (= 0.50 \times 10^3)$ - together with the rescaling of the exponent needed for subtraction - has made it too small (given only 2 digit arithmetic).

Then, in the calculation of x_1, nearly equal numbers were being subtracted, leading to catastrophic cancellation and gross error in x_1.

Pre-scaling and Pivoting

A first reaction might be to suspect that the problem illustrated above was caused by the relatively small size of the pivot element $a_{11} = 0.002$. This notion can be investigated further by first multiplying the first row by 1000 so that the augmented matrix is now

$$\begin{bmatrix} 2 & 1000 & | & 1000 \\ 1 & 1 & | & 2 \end{bmatrix}$$

and then repeating the calculation using 2 digit arithmetic as before. Again it will be seen that the error in the value found for x_1 is large.

It should be clear that error will be reduced if the *multipliers* are kept small. If necessary, a row beyond that containing the pivot can be exchanged with the pivot row to bring in the largest possible pivot element at that stage. This **pivoting** process is aided by **pre-scaling**, with elements in each row being divided by that having the largest magnitude in the row.

To pre-scale the pair of equations represented in matrix form by

$$\begin{bmatrix} 2 & -0.5 \\ 3 & -5 \end{bmatrix} \begin{bmatrix} x_1 \\ x_2 \end{bmatrix} = \begin{bmatrix} 4 \\ 2 \end{bmatrix}$$

we need to divide the first by 2 and the second by 5, so that they become

$$\begin{bmatrix} 1 & -0.25 \\ 0.6 & -1 \end{bmatrix} \begin{bmatrix} x_1 \\ x_2 \end{bmatrix} = \begin{bmatrix} 2 \\ 0.2 \end{bmatrix}.$$

For another example we return to the equations

$$\begin{bmatrix} 0.002 & 1 \\ 1 & 1 \end{bmatrix} \begin{bmatrix} x_1 \\ x_2 \end{bmatrix} = \begin{bmatrix} 1 \\ 2 \end{bmatrix}.$$

In this case the equations are already in scaled form (the largest coefficient in each equation is 1). The pivoting strategy requires that we exchange the first and second rows so that the pivot element (row 1, column 1 at this stage) is 1 rather than 0.002. Then the augmented matrix is

$$\begin{bmatrix} 1 & 1 & | & 2 \\ 0.002 & 1 & | & 1 \end{bmatrix}.$$

The application of Gaussian elimination using 2 digit arithmetic with chopping now results in $x_1 = 1$ and $x_2 = 1$, both much closer to the exact solution than the results found without pivoting.

For a wider discussion of pivoting strategies the reader should consult more advanced numerical analysis texts, such as that of Burden and Faires [1989], or Schwarz[1989] (and others cited therein). The main aim of the cursory study presented here has been to alert the reader to a few potential problems that arise from the limitations of computer representation of numbers.

4 Solution of Non-linear Equations

For many nonlinear equations it is possible to write down the solutions in closed form. A simple example is the quadratic equation $f(x) = x^2 - 2x - 3 = 0$. After factorizing the left-hand side, the equation becomes $(x-3)(x+1) = 0$, which has solutions $x = 3$ and $x = -1$.

However for the equation $x^2 - 2\sin(x) - 3 = 0$ we cannot obtain the solutions by simple manipulation, and must resort to other means. This usually involves making an initial estimate of the desired solution, followed by some iterative process which will improve the accuracy of the estimated value until the required accuracy is achieved.

For cases where $f(x)$ changes sign as x passes through the solution value the method of interval bisection will always converge, but the rate of convergence is slow, being linear. Another (usually) linear method that may often be used is that of fixed-point iteration. Much more rapid convergence is offered by Newton's method , and by the secant method.

These methods can be generalised to the case of several nonlinear equations, and this chapter includes examples of fixed-point iteration and Newton's method for more than one equation. After these, a modification of Newton's method (for one equation) in the case of a multiple solution is discussed.

Next, there is an example of an ill-conditioned nonlinear equation, and the chapter concludes with a topic that does not really belong to Numerical Analysis, namely the idea of a cobweb diagram. This concept is particularly useful to the discussion of fixed-point iteration, and is easily implemented on a spreadsheet.

§4.1   Solving Equations

Regarding the equation $x^2 - 2\sin(x) - 3 = 0$, we can at least establish that solutions do exist. This equation can be rewritten as $x^2 - 3 = 2\sin(x)$. Familiarity with the functions $y = x^2 - 3$ and $y = 2\sin(x)$, or the use of the graphing facility of a spreadsheet or some other software, indicates that their graphs do intersect. This is shown in Graph 4.1.1.

From Graph 4.1.1 it is evident that the equation $x^2 - 2\sin(x) - 3 = 0$ has two solutions, one near $x \approx -1.1$ and one near $x \approx 2.1$.

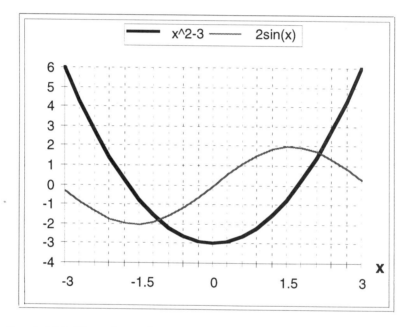

Graph 4.1.1 *Plots of* $y = x^2 - 3$ *and* $y = 2\sin(x)$, *showing two intersections.*

The use of mathematics in the modelling of real-life systems generally results in equations that cannot be solved "exactly", providing the motive for the various numerical methods introduced below.

4.1.1 Terminology:

A solution of an equation $f(x) = 0$ is also called a **root** of f, or a **zero** of f. We will be concerned here with finding the *real* solutions (if any) of equations. Some of the methods considered can be readily adapted to locate complex roots, however implementation on a spreadsheet (if possible) may be messy. We assume f to be continuous in some neighborhood of each of its roots.

4.1.2 Simple and multiple roots

Graphically, the (real) solutions (if any exist) of the equation $f(x) = 0$ are the values of x at which the graph of the function $f(x)$ crosses or touches the x-axis.

The graph will **cross** the x-axis in the case of a root of **odd** multiplicity (e.g. the root $x = 2$ of the equation $g(x) = (x-2)^3(x+3)^2 = 0$), and **touch** it for a root of **even** multiplicity (as for the root $x = -3$ of this equation). A root of (odd) multiplicity 1 is referred to as a **simple** root e.g. the root $x = 7$ of equation $(x-7)\tan(x) = 0$. This last equation has infinitely many roots, all of them simple.

At a (non-simple) root of odd multiplicity $f''(x)$ always changes sign, but near a root of even multiplicity $f''(x)$ has constant sign, and $f''(x)$ *may* change sign at a simple root.

These features are illustrated using $g(x) = (x-2)^3(x+3)^2 = 0$ in Graph 4.1.2. Root multiplicity is discussed further in §4.7.

The existence and approximate values of the roots can often be found by setting up a table of values of the function on a spreadsheet and then using its graphing facility. Once a root is roughly located its value can be found to higher accuracy by narrowing the range of x-values plotted (including the desired root, of course), using smaller step sizes.

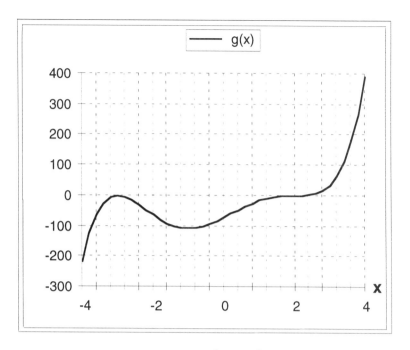

Graph 4.1.2 *The graph of $g(x) = (x-2)^3(x+3)^2$ touches the x-axis at x = -3, and crosses it at x = 2.*

4.1.3 Example

Consider the equation $f(x) = x - \exp(-x^2) = 0$. Tabulate and plot $f(x)$, initially over the interval $[-5,5]$, and then over narrower intervals to find more accurate estimates of the solution(s).

For a plot of $f(x)$ on $[-5,5]$, see Graph 4.1.3. In this case there is only one root, visible, lying between 0 and 1. In fact, since $f'(x) = 1 + 2x\exp(-x^2) > 0$ it follows that the equation has only the one solution.

Screen 4.1.1 shows a spreadsheet for the tabulation in column B of $f(x)$ for values of x (in column A) between 0 and 1, spaced at intervals of 0.1. It is clear that $f(x)$ changes sign (i.e. $f(x) = 0$ has a root) between $x = 0.6$ and $x = 0.7$. Changing the step-size (in cell D3) to 0.01 and the starting value (in cell B3) to 0.6 gives the more accurate estimate that the root lies between $x = 0.65$ and $x = 0.66$, and closer to the former value.

	A	B	C	D
1	Example 4.1.3	Tabulation of function		
2				
3	start at x=	=0	step-size =	=0.1
4	x	f(x)=x-exp(-x^2)		
5	=B$3	=A5-EXP(-A5^2)		
6	=A5+D$3	=A6-EXP(-A6^2)		

	A	B	C	D
1	Example 4.1.3	Tabulation of function		
2				
3	start at x=	0.000	step-size =	0.1
4	x	f(x)=x-exp(-x^2)		
5	0	-1.00		
6	0.1	-0.89		

Screen 4.1.1 *Tabulation of the function* $f(x) = x - \exp(-x^2)$.

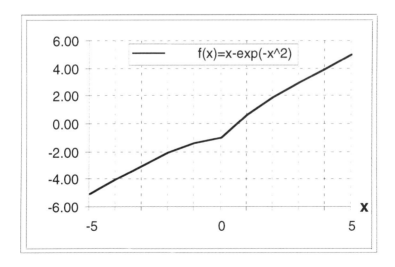

Graph 4.1.3 *A plot showing that* $f(x)$ *has a zero between 0 and 1.*

111

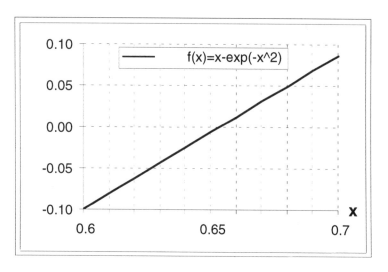

Graph 4.1.4 *Plotting over a narrower range shows the zero to be between 0.65 and 0.66.*

Graph 4.1.4 is generated from this latter set of data. This "manual" narrowing of the interval containing the root can be continued until the desired precision is achieved, but an automated version - some sort of iterative process - would be preferable. Perhaps the simplest such method for simple roots and roots of higher odd multiplicity is the **bisection method.**

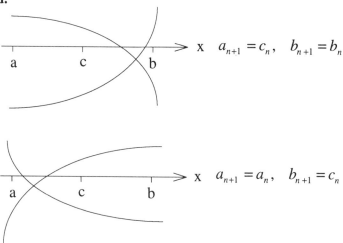

Fig. 4.2.1 *Four ways that $f(x)$ can change sign on [a,b], crossing the x-axis either to the left or to the right of the centre x = c.*

§4.2 The Interval Bisection method

Suppose we have an equation $f(x) = 0$ and that we know that the function f is continuous on an interval $[a,b]$, and that $f(a)$ and $f(b)$ have opposite signs, so that $f(a).f(b) < 0$. Then (by the Intermediate Value Theorem) there exists at least one value of x (x_0, say) with $a < x_0 < b$, such that $f(x_0) = 0$. In other words, the equation $f(x) = 0$ has a root x_0 between a and b.

With regard to *where* it crosses the x-axis, the graph of $f(x)$ must behave as in one of the four cases depicted in Fig. 4.2.1. Of course the shape of the graph may differ from that shown, and we are assuming that $f(c) \neq 0$.

In practice we may have found the values of a and b by the sort of graphing exercise outlined above. We then proceed as follows:

Compute the value of $f(c)$ at the midpoint $c = \frac{1}{2}(a+b)$ of the interval $[a,b]$. Then **either**

case (1): $f(a) < 0$. If $f(c) < 0$ replace a with c. If $f(c) > 0$ replace b with c, **or**

case (2): $f(a) > 0$ do the opposite i.e. if $f(c) > 0$ replace a with c and if $f(c) < 0$ replace b with c.

The root now lies in a smaller interval $[a,b]$. Repeat these steps until desired accuracy is reached.

Of course if $f(c) = 0$ we know (exactly, or at least to within the smallest number that the computer can represent) that $x = c$ is a root!

Each successive application of this algorithm **halves** the length of the interval containing the root. With $a_0 = a$ and $b_0 = b$ we get a sequence of intervals $[a_0, b_0], [a_1, b_1], [a_2, b_2], [a_3, b_3], \ldots$, which we terminate when the length of the interval is sufficiently small. If we let $c_n = \frac{1}{2}(a_n + b_n)$, the interval bisection algorithm can be simply stated as:

If $f(a_n)$ and $f(c_n)$ have the same sign, let $a_{n+1} = c_n$ and $b_{n+1} = b_n$, but if $f(a_n)$ and $f(c_n)$ have opposite sign, let $a_{n+1} = a_n$ and $b_{n+1} = c_n$

The different cases are also summarised in Fig. 4.2.1.

Notes: (1) This method cannot be used for roots having even multiplicity since in that case f(x) does not change sign at the root. (2) It is important to ensure that there is only *one* root in the initial interval.

4.2.1 Example

Apply the bisection method to the problem of Example 4.1.3, $f(x) = x - \exp(-x^2) = 0$, performing 10 iterations with the initial interval $[0.6, 0.7]$ counting the selection of the initial interval as the first (with $n = 0$).

See screen 4.2.1. The formulas in cells B5 and C5 (and below them) calculate the a_n and b_n values as prescribed in the box above, with the starting values in cells B4 and C4. After 10 iterations the root lies between $x = 0.6527344$ and $x = 0.6529297$, i.e. we know it to 3 decimal places. A further 10 iterations improves this to $[0.65291862, 0.65291881]$, i.e. to 6 d.p.

A drawback of the bisection method is that it converges slowly - a large number of iterations may be needed to reach the desired accuracy. On the positive side, it always will converge and is often used as a way of providing a starting value for other faster methods. An **upper bound** on the error after n iterations is given by the following theorem:

4.2.2 Theorem

$$|c_n - x_0| \le \frac{b - a}{2^n}, \quad n \ge 1$$

Exercise: Prove this result.

Note: For the example above, after 10 iterations ($n = 9$) the error is *at most* $(0.7 - 0.6) = 0.1 / 512 \approx 0.0002$. The *actual* error is (approximately) $0.652918 - 0.652832 = 0.000086$, quite a bit smaller.

	A	B	C
1	Example 4.2.1		
2	Interval Bisection		
3	n	a(n)	b(n)
4	=0	=0.6	=0.7
5	=A4+1	=IF(E4*F4>=0,D4,B4)	=IF(E4*F4<0,D4,C4)

	D	E	F
3	c(n)	f(a(n))	f(c(n))
4	=(B4+C4)/2	=B4-EXP(-B4^2)	=D4-EXP(-D4^2)

	A	B	C	D	E	F
1	Example 4.2.1					
2	Interval Bisection					
3	n	a(n)	b(n)	c(n)	f(a(n))	f(c(n))
4	0	0.6	0.7	0.65	-1E-01	-5E-03
5	1	0.65	0.7	0.675	-5E-03	4E-02
6	2	0.65	0.675	0.6625	-5E-03	2E-02
7	3	0.65	0.6625	0.65625	-5E-03	6E-03
8	4	0.65	0.65625	0.653125	-5E-03	4E-04
9	5	0.65	0.653125	0.6515625	-5E-03	-3E-03
10	6	0.6515625	0.653125	0.65234375	-3E-03	-1E-03
11	7	0.6523438	0.653125	0.652734375	-1E-03	-3E-04
12	8	0.6527344	0.653125	0.652929688	-3E-04	2E-05
13	9	0.6527344	0.6529297	0.652832031	-3E-04	-2E-04

Screen 4.2.1 Solution of $f(x) = x - \exp(-x^2) = 0$ using interval bisection.

Exercises 4.2

(1) Use the bisection method and the equation $x^5 - 7 = 0$ to find $\sqrt[5]{7}$ with error less than 10^{-6}. Compare the actual error and the upper bound for the error.

(2) The equation $f(x) = 1 + e^{-x} - 2\sin(\pi x) = 0$ has infinitely many roots. Use the bisection method to find the first few positive roots, noting the effect of making the initial interval too large.

(3) Rewrite the bisection algorithm so that it tests the sign of the product $f(b)f(c)$ instead of $f(a)f(c)$. Test your new algorithm by repeating the preceding two exercises, using it in place of the previous algorithm.

§4.3 Solving an Equation using Fixed-Point Iteration

Assume that we can express our equation $f(x) = 0$ in the alternative form $g(x) = x$. For example, $f(x) = x^2 + x - 1 = 0$ can be written as $g(x) = \sqrt{1-x} = x$, or as $g(x) = 1 - x^2 = x$.

Note: The solutions of $g(x) = x$ may not include *all* of the solutions of $f(x) = 0$. This is the case in the example above, where $g(x) = \sqrt{1-x} = x$ only gives the positive root of $f(x) = x^2 + x - 1 = 0$. The other (negative) root is the solution of $-\sqrt{1-x} = x$.

A solution to the equation $g(x) = x$ is called a **fixed point** of g.

Thus the problem of finding some or all of the roots of the equation $f(x) = 0$ is the same problem as that of finding the fixed points of $g(x)$.

If $g(x)$ satisfies certain conditions in a neighborhood of s then, given a sufficiently good initial approximation x_0 to a fixed point $x = s$ of a function $g(x)$, we can generate a sequence $x_0, x_1, x_2, \ldots$ of approximations convergent to s, using the **fixed point iteration formula**

$$x_n = g(x_{n-1}), \quad n \geq 1.$$

If the sequence converges to s and $g(x)$ is continuous then $s = g(s)$ and we have a solution s of $x = g(x)$.

Sufficient conditions for the existence and uniqueness of a fixed point are given in the following theorem:

4.3.1 Fixed Point (Existence) Theorem

Let the function g be continuous on the interval $[a,b]$, with (i) $a \leq g(x) \leq b$ for all x in $[a,b]$. Then g has a fixed point in $[a,b]$.

Also, if (ii) $g'(x)$ exists for all x in (a,b) and there exists a positive constant $k < 1$ such that $|g'(x)| \leq k < 1$ for all x in (a,b), then g has a *unique* fixed point in $[a,b]$.

An example where such a convergent sequence of iterations is obtained is shown in Fig. 4.3.1. The zig-zag line with arrows tracing the iteration is called a **cobweb diagram.** These diagrams are discussed further in §4.9.

Naively, the theorem above says that $g(x)$ has a fixed point in $[a,b]$ if the graph of $y = g(x)$ intersects that of $y = x$ in $[a,b]$. If, in addition, $|g'(x)| \le k < 1$ in $[a,b]$, then only one intersection is possible. A couple of possible cases are shown in Fig. 4.3.2.

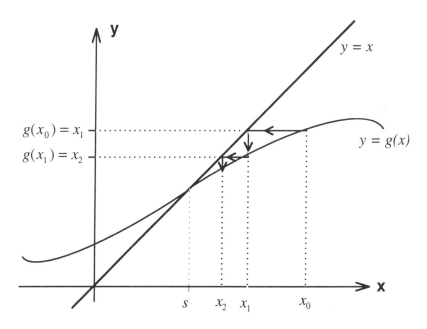

*Fig. **4.3.1** A sequence of iterates convergent to the fixed point at x = s. The arrowed zig-zag line tracing the iteration is called the cobweb diagram for this iteration.*

Exercises 4.3.1

(1) The conditions of Theorem 4.3.1 are *sufficient* but not *necessary* for the existence of a *unique* fixed point. Verify this by considering the existence the positive root of $x^2 - 4x - 1 = 0$ as a fixed point of $g(x) = \frac{1}{4}(x^2 - 1)$, lying in the interval [4,5].

(**Hint**: Calculate $g(4)$ and $g'(5)$, for example)

(2) For the cubic function $f(x) = x^3 + 3x^2 - 9$, show that the equation $f(x) = 0$ has a root s in the interval [1, 2]. This equation can be written in the form $g(x) = x$ in (at least) 5 ways:

$$g_1(x) = 9 + x - 3x^2 - x^3$$

$$g_2(x) = \sqrt{\frac{9}{x} - 3x}$$

$$g_3(x) = \sqrt{\frac{9 - x^3}{3}}$$

$$g_4(x) = \sqrt{\frac{9}{x+3}}$$

$$g_5(x) = \frac{2x^3 + 3x^2 + 9}{3x^2 + 6x}$$

Construct a spreadsheet containing all these iterations (for help see Screen 4.3.1) and reconcile the results with Theorem 4.3.3 (below). You may find it useful to use a spreadsheet to plot a graph of the derivative $g'(x)$ in each case.

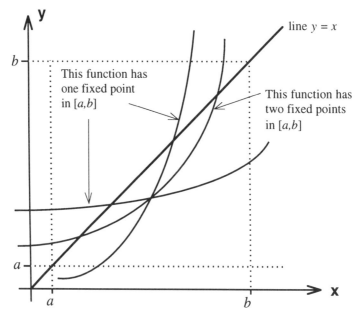

Fig. 4.3.2 *Illustrating existence of fixed points.*

4.3.2 Example

Apply fixed-point iteration to find the positive root of $x^2 + x - 1 = 0$.

This equation can be expressed in suitable form in at least two ways: (a) as $g(x) = 1 - x^2 = x$, and (b) as $h(x) = \sqrt{1-x} = x$. The required (positive) root is a fixed point of both $g(x)$ and $h(x)$. (The *other* root of $x^2 + x - 1 = 0$ is fixed point of both $g(x)$ and $l(x) = -\sqrt{1-x}$.)

Screen 4.3.1 shows both iterations (in columns B and D), using the starting value $x_0 = 0.5$. The $g(x)$ iteration does not converge - in fact it settles down to a period = 2 oscillation, whereas the $h(x)$ iteration (encountered earlier in chapter 2, see Example 2.4.2) does appear to converge to the desired root. With $s = \frac{1}{2}(-1 + \sqrt{5})$ we find $|g'(s)| = 2s = -1 + \sqrt{5} > 1$ and $|h'(s)| = \frac{1}{2}(1 - s)^{-1} = [2(3 - \sqrt{5})]^{-\frac{1}{2}} < 1$. The observed behaviour of these two cases can now be understood in view of Theorem 4.3.3.

	A	B	C
1	Example 4.3.2		
2		Fixed pt iteration	to solve
3		g(x)=1-x^2	
4	n	x(n)=g(x(n-1))	|error|
5	=0	=0.5	=ABS(E$1-B5)
6	=A5+1	=1-B5^2	=ABS(E$1-B6)

	D	E
1	exact root=	=(-1+SQRT(5))/2
2	x^2+x-1=0	
3	h(x)=sqrt(1-x)	
4	x(n)=h(x(n-1))	|error|
5	=0.5	=ABS(E$1-D5)
6	=SQRT(1-D5)	=ABS(E$1-D6)

Screen 4.3.1. *Two different fixed point iterations to find the positive root of* $x^2 + x - 1 = 0$.

	A	B	C	D	E
1	Example 4.3.2			exact root=	0.618034
2		Fixed pt iteration	to solve	x^2+x-1=0	
3		g(x)=1-x^2		h(x)=sqrt(1-x)	
4	n	x(n)=g(x(n-1))	|error|	x(n)=h(x(n-1))	|error|
5	0	0.500000	1.2E-01	0.500000	1.2E-01
6	1	0.750000	1.3E-01	0.707107	8.9E-02
7	2	0.437500	1.8E-01	0.541196	7.7E-02

Screen 4.3.1. *(continued)*

4.3.3 Fixed-Point (Convergence)Theorem

Suppose the function $g(x)$ satisfies conditions (i) and (ii) of Theorem 4.3.1. Then if x_0 is any number in $[a,b]$ the sequence defined by $x_n = g(x_{n-1})$, $n \geq 1$ converges to the unique fixed point s in $[a,b]$.

If $|g'(x)| > 1$ on all intervals containing s, the iteration does not converge to s.

Note: Condition (ii) is not necessary for the *convergence* of the fixed point iteration. For example, suppose that $|g'(s)| = 1$ and $g''(s) < 0$, as depicted in Fig. 4.4. In this case the iteration is convergent to s if $x_0 > s$ (and $|g'(x)| < 1$) but not if $x_0 < s$ (and $|g'(x)| > 1$), assuming x_0 to be sufficiently close to s.

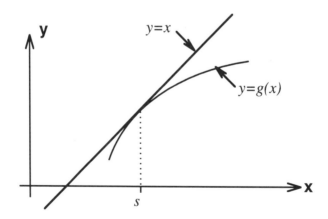

Fig. 4.3.3 *For this function, an iteration starting with $x > s$ (but not too large) will converge to s, but one starting with $x < s$ will diverge away from s.*

Returning to Example 4.3.2, condition (i) is satisfied on the interval $[0,1]$ by both functions. However $|g'(x)| < 1$ only if $|x| < \frac{1}{2}$ so that (since $s > \frac{1}{2}$) there is no interval *containing* s over which $|g'(x)| < 1$. On the other hand, $|h'(x)| < 1$ if $x < \frac{3}{4}$, and so (for example) $[0, 0.75]$ is a suitable interval here.

Another observation is that for the convergent case (b) the number of iterations required to reach 5 d.p. accuracy is comparable with that for the bisection method. Theorem 4.3.3 leads to the following further information regarding the error and the rate of convergence:

4.3.4 Corollary $|x_n - s| \leq k^n \max\{x_0 - a, x_0 - b\}, \quad n \geq 1$

4.3.5 Corollary $|x_n - s| \leq k^n \dfrac{|x_0 - x_1|}{1 - k}, \quad n \geq 1$

i.e. the smaller is the value of k, the more rapid is the convergence.

Exercise 4.3.2

Apply the fixed-point theorem to the quadratic map $g(x) = \mu x(1 - x)$ that was studied in Exercise 2.6(5). The position of the fixed point and the convergence or otherwise of the iteration will both depend on μ. Compare the theoretical predictions with the observed behaviour.

4.3.6 Theorem   Rate of Convergence of Fixed-point Iteration

When convergent, the sequence defined by a fixed-point iteration (when $g'(s) \neq 0$) converges only **linearly** to the unique fixed point s in $[a, b]$.

Faster convergence of fixed-point iteration can be achieved using an acceleration method from chapter 2, as demonstrated in Example 2.4.2. Note that Steffenson's method gives quadratic convergence without the need for a derivative, a point that will be appreciated when (later) the Newton-Raphson and secant methods have been discussed.

We now seek methods which have a higher order rate of convergence. Theorem 4.3.6 implies that a fixed-point iteration will only give a higher rate when $g'(s) = 0$. This is covered in the next theorem.

4.3.7 Theorem

Suppose s is a solution of the equation $x = g(x)$, that $g'(s) = 0$ and that g'' is continuous and strictly bounded on (a, b) (with $a < s < b$). Then, for x_0 sufficiently close to s, the sequence defined by $x_n = g(x_{n-1})$, $n \geq 1$, converges **at least quadratically** to s.

Theorem 4.3.7 leads directly to Newton's method - which we will approach from a different direction.

Exercise 4.3.3

The equation $x^2 - (3 + 2a)x + 2 + 2a + a^2 = 0$ can be rewritten as

$$g(x) = (x - 1 - a)^2 + 1 = x$$

Construct a spreadsheet for the fixed point iteration defined by $g(x)$ and observe the behaviour of the iteration for $0 \leq a \leq \frac{1}{2}$, using $x_0 = 1.2$ to obtain convergence to the smaller of the two positive roots, viz.

$$s = \tfrac{1}{2}[3 + 2a - \sqrt{1 + 4a}]$$

Include a column showing the value of the derivative $g'(x) = 2(x - 1 - a)$, and another to estimate the order of convergence (as practiced in Exercise 2.2.3(2)). Experiment with the value of x_0.

The other root is

$$s = \tfrac{1}{2}[3 + 2a + \sqrt{1 + 4a}]$$

Can the iteration discussed above give this root? Why?

§4.4 The Newton-Raphson Method

Newton's method (when it works) is a powerful way of solving $f(x) = 0$. It can be derived in several ways, the most transparent being geometrically motivated, as depicted in Fig. 4.4.1.

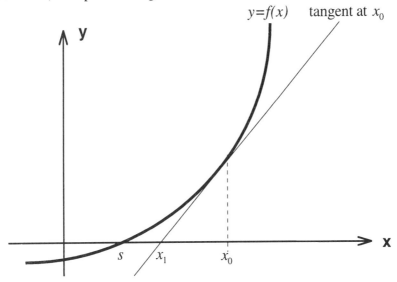

Fig. 4.4.1 *The geometric basis for Newton's method: replace the curve with its tangent.*

Let us assume that x_0 lies sufficiently close to a root s. On the graph of $y = f(x)$ construct the tangent line at the point $(x_0, f(x_0))$ and take the point where this line cuts the x-axis to be the next iterate x_1. In other words, we will estimate the value of s by replacing the curve with its tangent at the initial "guess" value x_0.

A little coordinate geometry gives $x_1 = g(x_0) = x_0 - \dfrac{f(x_0)}{f'(x_0)}$.

Using x_1 in place of x_0 leads to the next estimate

$$x_2 = g(x_1) = x_1 - \frac{f(x_1)}{f'(x_1)}$$

from which we deduce the iteration formula known as Newton's method:

Newton's method

$$x_n = g(x_{n-1}) = x_{n-1} - \frac{f(x_{n-1})}{f'(x_{n-1})}, \quad n \geq 1 \qquad (4.4.1)$$

This method will fail if $f'(x_{n-1}) = 0$ for some n. In fact it is best when $f'(x)$ is bounded away from zero near $x = s$, in which case the convergence is quadratic. We will return later to the case where $f'(s) = 0$ (which implies that $x = s$ is not a simple root).

4.4.1 Example

Apply Newton's method to the problem of Example 4.3.2, finding the solutions of $f(x) = x^2 + x - 1 = 0$.

See Screen 4.4.1. The iteration sequence (column B) in this case has

$$g(x) = x - \frac{f(x)}{f'(x)} = x - \frac{x^2 + x - 1}{2x + 1}$$

With the initial value $x_0 = 0.5$ we have better than 3 d.p. accuracy after 2 iterations, and about as good as machine precision after only 5 iterations!

	A	B	C
1	Example 4.4.1	exact root =	=(-1+SQRT(5))/2
2		Newton's method	
3		equation f(x)=	x^2+x-1=0
4	n	x(n)	|error|
5	=0	=0.5	=ABS(B5-C$1)
6	=A5+1	=B5-(B5^2+B5-1)/(2*B5+1)	=ABS(B6-C$1)

Screen 4.4.1 Use of Newton's method to solve $x^2 + x - 1 = 0$.

	A	B	C
1	Example 4.4.1	exact root =	0.618033989
2		Newton's method	
3		equation f(x) =	x^2+x-1=0
4	n	x(n)	|error|
5	0	0.5	1.180E-01
6	1	0.625	6.966E-03
7	2	0.618055556	2.157E-05
8	3	0.618033989	2.080E-10
9	4	0.618033989	1.110E-16
10	5	0.618033989	1.110E-16

Screen 4.4.1 (*continued*)

Exercises 4.4

1) Experiment with the spreadsheet of Example 4.4.1 by varying the value of x_0, e.g. starting with $x_0 = 1.0$ and reducing it by (say) 0.1 and noting the apparent limit of the sequence, particularly as x_0 passes through -0.5. How would you interpret the observed behaviour?

(2) Explore the application of Newton's method to finding the root of

$$f(x) = \frac{5x - 4}{x - 1}$$

(3) Repeat (2) with the function $f(x) = x - 2\tan(x)$. You may find it helpful to tabulate and graph this function first in order to see roughly where its roots lie.

Note: The preceding exercises illustrate the fact that *sometimes* Newton's method is relatively insensitive to x_0, and also that using a value of x_0 sufficiently removed from the desired root may result in the computation of a different root (Fig 4.4.2), or even a divergent sequence (Fig 4.4.3).

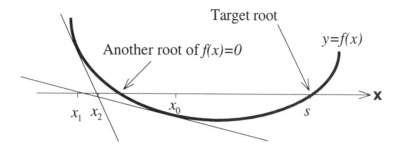

Fig. 4.4.2 *If the starting value is too far from the desired root s, Newton's method may give convergence to another root...*

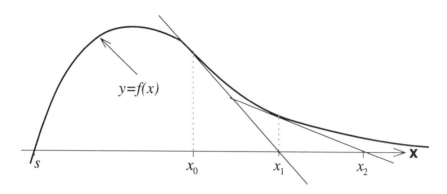

Fig. 4.4.3 *... or it may lead to a divergent sequence.*

An alternative derivation of Newton's method uses the Taylor series for $f(x)$, expanding about the point $x = s$.

Assuming that $f''(x)$ exists on $[a,b]$ and that $|x_0 - s|$ is small, with $f'(x_0) \neq 0$, Taylor's Theorem gives

$$f(x) = f(x_0) + (x - x_0)f'(x_0) + \tfrac{1}{2}(x - x_0)^2 f''(\xi)$$

where ξ lies between x and x_0. With $x = s$, noting that $f(s) = 0$, this gives

$$0 = f(x_0) + (s - x_0)f'(x_0) + \tfrac{1}{2}(s - x_0)^2 f''(\xi)$$

Now, if $|x_0 - s|$ and $f''(x)$ are small then the third term on the right-hand side is negligible and thus

$$0 \approx f(x_0) + (s - x_0)f'(x_0) \text{ i.e. } s \approx x_0 - \frac{f(x_0)}{f'(x_0)}$$

which is the basis for Newton's method. This derivation emphasises the importance of selecting an initial value that is sufficiently close to the desired root. Intuitively, if $f(x)$ has large curvature (i.e. large f'') near s, the tangent may well intersect the x-axis closer to another root if $|x_0 - s|$ is too large.

§4.5   The Secant Method

A disadvantage of Newton's method is the need to evaluate the derivative $f'(x_n)$ at each step, which may not be convenient for many reasons. For example, if $f(x_n)$ is the result of other numerical computations instead of the evaluation of an explicit function then $f'(x_n)$ is not directly available.

For some functions, the expression for $f'(x_n)$ is more complex than that for $f(x_n)$, requiring considerably more arithmetic calculations (i.e. computer time) than $f(x_n)$. An example is the function

$$f(x) = \frac{x^3}{e^x - 1} \Rightarrow f'(x) = \frac{x^2(3-x)e^x - 3x^2}{(e^x - 1)^2}.$$

In a later chapter we derive a finite-difference approximation for the first derivative,

$$f'(x_n) \approx \frac{f(x_n) - f(x_{n-1})}{x_n - x_{n-1}}$$

and when this is substituted into the Newton formula, equation 4.4.1, we get the formula for the **secant method**:

Secant method

$$x_n = x_{n-1} - \frac{f(x_{n-1})(x_{n-1} - x_{n-2})}{f(x_{n-1}) - f(x_{n-2})}, \quad n = 2, 3, 4, \ldots \qquad (4.5.1)$$

Of course we need *two* initial guesses x_0 and x_1 to use the secant method.

The rate of convergence is not quite quadratic, with order $\alpha = (1 + \sqrt{5})/2 \approx 1.62$.

Geometrically (see Fig 4.5.1) we are using the chord joining the points on the graph of $f(x)$ corresponding to x_0 and x_1, taking its intersection with the x-axis as the improved root x_2.

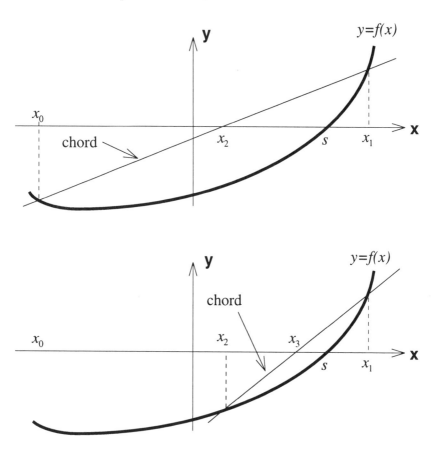

Fig. 4.5.1 *Geometrical basis for the secant method: replace the curve with a chord.*

4.5.1 Example

We solve again the problem of Example 4.3.2, the equation

$$f(x) = x^2 + x - 1 = 0,$$

this time using the secant method with initial guesses 0.5 and 0.6.

See Screen 4.5.1. Note the rapid convergence of the secant sequence in column B. There is just one function evaluation per step, in column C. Eventually a "divide-by-zero" error alert occurs when successive function values differ by less than machine precision and the denominator is set to zero.

	A	B	C	D
1	Example	exact root =	=(-1+SQRT(5))/2	
2	4.5.1	Secant method		
3		equation f(x)=	x^2+x-1=0	
4	n	x(n)	f(x(n))	|error|
5	=0	=0.5	=B5^2+B5-1	=ABS(B5-C$1)
6	=A5+1	=0.6	=B6^2+B6-1	=ABS(B6-C$1)
7	=A6+1	=B6-C6*(B6-B5)/(C6-C5)	=B7^2+B7-1	=ABS(B7-C$1)

	A	B	C	D
1	Example	exact root =	0.618033989	
2	4.5.1	Secant method		
3		equation f(x)=	x^2+x-1=0	
4	n	x(n)	f(x(n))	|error|
5	0	0.5	-2.500E-01	1.18E-01
6	1	0.6	-4.000E-02	1.80E-02
7	2	0.619047619	2.268E-03	1.01E-03
8	3	0.618025751	-1.842E-05	8.24E-06
9	4	0.618033985	-8.346E-09	3.73E-09
10	5	0.618033989	3.064E-14	1.37E-14
11	6	0.618033989	0.000E+00	0.00E+00
12	7	0.618033989	0.000E+00	0.00E+00
13	8	#DIV/0!	#DIV/0!	#DIV/0!

Screen 4.5.1 *Finding the roots of $x^2 + x - 1 = 0$ using the secant method.*

Exercises 4.5

(1) Experiment with the initial values used in Example 4.5.1 above.

(2) Repeat Exercise 4.4(2) using the secant method. What restrictions must be applied to the choice of the initial guesses?

(3) Use the secant method to find the positive root of the equation

$$\frac{3-x}{x^2+1} - \frac{e^x}{x^2+2} = 0$$

Derive the Newton iteration formula for this problem in order to see how complex it is compared with the secant method.

(4) The **regula falsi** method generates a sequence of intervals $[a_i, b_i]$ that all contain the desired root of $f(x) = 0$, and it assumes that $f(x)$ changes sign at that root. Let c_n be the x-intercept of the chord joining $[a_n, f(a_n)]$ to $[b_n, f(b_n)]$. The process is stopped when $f(c_n)$ is sufficiently small, or the next interval $[a_{n+1}, b_{n+1}]$ is given by

$$a_{n+1} = c_n, b_{n+1} = b_n \text{ if } f(a_n)f(c_n) > 0,$$

or

$$a_{n+1} = a_n, b_{n+1} = c_n \text{ if } f(a_n)f(c_n) < 0$$

with

$$c_n = \frac{a_n f(b_n) - b_n f(a_n)}{f(b_n) - f(a_n)}$$

Apply the regula falsi method to the preceding exercises.

(5) Newton's method can be used without recalculating the derivative at every step, using instead the value found for $f'(x_0)$ in all subsequent steps. The convergence is not quadratic, however, but only linear. Use this variant to solve the problems of Exercises 4.4.

§4.6* Systems of Several Nonlinear Equations

We consider next the matter of solving N simultaneous nonlinear equations in N variables, which have the form

$$f_1(x_1, x_2, \cdots x_N) = 0$$
$$f_2(x_1, x_2, \cdots x_N) = 0$$
$$\vdots$$
$$f_N(x_1, x_2, \cdots x_N) = 0$$

These are conveniently written in vector form as

$$\mathbf{F}(\mathbf{x}) = \mathbf{0}, \text{ where } \mathbf{x} = (x_1, x_2, \cdots, x_N)$$

and

$$\mathbf{F}(\mathbf{x}) = (f_1(\mathbf{x}), f_2(\mathbf{x}), \cdots, f_N(\mathbf{x})).$$

Some of the methods employed for the case of one equation are readily adapted to this problem, for instance fixed-point iteration and Newton's method.

Example

Consider the system of equations

$$0 = f_1(x_1, x_2, x_3) = 4x_1 - \cos(x_1 x_2) - 1$$
$$0 = f_2(x_1, x_2, x_3) = 1 - 4x_2 - \tfrac{1}{2}e^{-x_2 x_3}$$
$$0 = f_3(x_1, x_2, x_3) = 1 - 4x_3 - x_1 x_2 x_3$$

In order to define a fixed-point iteration they may be written as

$$x_1 = g_1(x_1, x_2, x_3) = \tfrac{1}{4}(1 + \cos(x_1 x_2))$$
$$x_2 = g_2(x_1, x_2, x_3) = \tfrac{1}{4}(1 - \tfrac{1}{2}e^{-x_2 x_3}) \qquad (4.6.1)$$
$$x_3 = g_3(x_1, x_2, x_3) = \tfrac{1}{4}(1 - x_1 x_2 x_3)$$

4.6.1 Fixed Points of Functions of Several Variables

Assume that the equations $\mathbf{F}(\mathbf{x}) = \mathbf{0}$ can be written in the form $\mathbf{G}(\mathbf{x}) = \mathbf{x}$. The point $\mathbf{p}$ is a **fixed point** of $\mathbf{G}$ if $\mathbf{G}(\mathbf{p}) = \mathbf{p}$.

The following theorem is similar to Theorem 4.3.3 (the Fixed Point Convergence Theorem), giving sufficient conditions for the existence of a fixed point $\mathbf{p}$ of $\mathbf{G}$ and of an iteration that converges to $\mathbf{p}$.

Notation: The kth iterate is $\mathbf{x}^{(k)} = (x_1^{(k)}, x_2^{(k)}, x_3^{(k)}, ..., x_N^{(k)})$.

4.6.2 Theorem

Let $D = \{(x_1, x_2, \cdots x_N) \mid a_i \leq x_i \leq b_i , i = 1, 2, ..., N\} \subset R^N$ for some constants a_i and $b_i, i = 1, 2, ..., N$ Let $\mathbf{G}(\mathbf{x})$ be a continuous function from D into R^N such that $\mathbf{x} \in D \Rightarrow \mathbf{G}(\mathbf{x}) \in D$. Then $\mathbf{G}$ has a fixed point in D.

Further, if $\mathbf{G}$ has continuous partial derivatives and there exists a positive constant $K < 1$ with

$$\left| \frac{\partial g_i(\mathbf{x})}{\partial x_j} \right| \leq \frac{K}{N} \text{ when } \mathbf{x} \in D, \, i, j = 1, 2, ...N$$

then (with $\mathbf{x}^{(0)} \in D$) the sequence $\mathbf{x}^{(0)}, \mathbf{x}^{(1)}, \mathbf{x}^{(2)}, ...$ defined by

$$\mathbf{x}^{(k)} = \mathbf{G}(\mathbf{x}^{(k-1)}) \text{ for each } k \geq 1$$

converges to a unique fixed point $\mathbf{p}$ in D.

4.6.3 Example

Apply fixed point iteration to the 3 equations 4.6.1 above.

Let
$$D = \{(x_1, x_2, x_3) \mid 0 \le x_1 \le 1, \quad 0 \le x_2 \le 1, \quad 0 \le x_3 \le 1\} \subset R^3$$

and
$$G(\mathbf{x}) = (g_1(\mathbf{x}), g_2(\mathbf{x}), g_3(\mathbf{x})), \text{ where } \mathbf{x} = (x_1, x_2, x_3). \text{ Then}$$

$$0 \le g_1(x_1, x_2, x_3) \le \tfrac{1}{2}, \quad \tfrac{1}{8} \le g_2(x_1, x_2, x_3) < \tfrac{1}{4}, \quad 0 \le g_3(x_1, x_2, x_3) \le \tfrac{1}{4}$$

So $G(\mathbf{x})$ maps D into itself and thus G has a fixed point in D. Further, considering the partial derivatives,

$$\left| \frac{\partial g_1}{\partial x_1} \right| = \left| -\tfrac{1}{4} x_2 \sin(x_1 x_2) \right| < \tfrac{1}{4} \sin(1) < \tfrac{1}{4},$$

$$\left| \frac{\partial g_1}{\partial x_2} \right| = \left| -\tfrac{1}{4} x_1 \sin(x_1 x_2) \right| < \tfrac{1}{4} \sin(1) < \tfrac{1}{4},$$

$$\left| \frac{\partial g_1}{\partial x_3} \right| = 0, \quad \left| \frac{\partial g_2}{\partial x_1} \right| = 0, \quad \left| \frac{\partial g_2}{\partial x_2} \right| = \left| \tfrac{1}{8} x_3 e^{-x_2 x_3} \right| \le \tfrac{1}{8}, \quad \left| \frac{\partial g_2}{\partial x_3} \right| = \left| \tfrac{1}{8} x_2 e^{-x_2 x_3} \right| \le \tfrac{1}{8}$$

$$\left| \frac{\partial g_3}{\partial x_1} \right| = \left| -\tfrac{1}{4} x_2 x_3 \right| \le \tfrac{1}{4}, \quad \left| \frac{\partial g_3}{\partial x_2} \right| = \left| -\tfrac{1}{4} x_1 x_3 \right| \le \tfrac{1}{4}, \quad \left| \frac{\partial g_3}{\partial x_3} \right| = \left| -\tfrac{1}{4} x_1 x_2 \right| \le \tfrac{1}{4}$$

Thus we have
$$\left| \frac{\partial g_i}{\partial x_j} \right| \le \tfrac{1}{4} = \tfrac{K}{3} \text{ for } i, j = 1, 2, 3 \text{ and where } K = \tfrac{3}{4} < 1 .$$

Hence the iteration $\mathbf{x}_{n+1} = G(\mathbf{x}_n)$ with $\mathbf{x}_0 \in D$ will converge to a fixed point in D. The spreadsheet in Screen 4.6.1 shows this sequence, with the three variables (as given in equations 4.6.1) placed in columns B, C, and D.

The solution is (0.49948, 0.12890, 0.24604), to 5 d.p.

	A	B	C	D
1	Example	3D Fixed Point	Iteration	
2	4.6.3			
3	n	x1(n)	x2(n)	x3(n)
4	0	=0.5	=0.5	=0.5
5	=A4+1	=(1+COS(B4*C4))/4	=(1-EXP(-C4*D4)/2)/4	=(1-B4*C4*D4)/4

	A	B	C	D
1	Example	3D Fixed Point	Iteration	
2	4.6.3			
3	n	x1(n)	x2(n)	x3(n)
4	0	0.5	0.5	0.5
5	1	0.492228105	0.152649902	0.21875
6	2	0.499294606	0.1291051	0.245890859

Screen 4.6.1 *A system of 3 nonlinear equations solved by fixed point iteration.*

4.6.4 Newton's Method for Several Variables

Earlier we noted that for one equation $f(x) = 0$, Newton's method resulted from finding the intersection at $x = x_n$ of the tangent line to $f(x)$ at x_{n-1}, with the x-axis.

For two equations $f(x, y) = 0$ and $g(x, y) = 0$, each equation defines a curve in the xy plane, and the exact solutions correspond to the intersections of these curves, if any. The natural extension of Newton's method to two equations is as follows:

For each of the two surfaces $z = f(x, y)$ and $z = g(x, y)$ construct the tangent plane corresponding to the point (x_{n-1}, y_{n-1}) on the respective surface. Each of these tangent planes cuts the xy plane in a straight line. Now (assuming that the planes are not parallel) let the point of intersection of these two lines be the next iterate, (x_n, y_n).

This construction results in the following formulae (which may also be derived from the Taylor series):

$$\begin{bmatrix} x_n \\ y_n \end{bmatrix} = \begin{bmatrix} x_{n-1} \\ y_{n-1} \end{bmatrix} - \mathbf{J}^{-1} \begin{bmatrix} f(x_{n-1}, y_{n-1}) \\ g(x_{n-1}, y_{n-1}) \end{bmatrix}$$

Here, $\mathbf{J}$ is the Jacobian matrix

$$\mathbf{J} = \begin{bmatrix} f_x & f_y \\ g_x & g_y \end{bmatrix}, \text{ where } f_x = \frac{\partial f}{\partial x}, f_y = \frac{\partial f}{\partial y}, g_x = \frac{\partial g}{\partial x}, g_y = \frac{\partial g}{\partial y},$$

and all the partial derivatives are evaluated at (x_{n-1}, y_{n-1}). A generalization to several equations is now obvious,

$$\mathbf{x}^{(n)} = \mathbf{x}^{(n-1)} - \mathbf{J}^{-1}\mathbf{F}(\mathbf{x}^{(n-1)}),$$

but let us examine the specific formulae for the case of two equations. These are:

$$x_n = x_{n-1} - \frac{g_y f(x_{n-1}, y_{n-1}) - f_y g(x_{n-1}, y_{n-1})}{f_x g_y - f_y g_x}$$

and   (4.6.2)

$$y_n = y_{n-1} - \frac{f_x g(x_{n-1}, y_{n-1}) - g_x f(x_{n-1}, y_{n-1})}{f_x g_y - f_y g_x}$$

For the case of two equations, we have the **Jacobian**

$$\det \mathbf{J} = f_x g_y - f_y g_x = J(x, y), \text{ say.}$$

Note: One computational difficulty with Newton's method for several variables is that it requires, at each step, the solution (for $\mathbf{x}^{(n)}$) of the system of linear equations $\mathbf{Jx}^{(n)} = \mathbf{Jx}^{(n-1)} - \mathbf{F}(\mathbf{x}^{(n-1)})$. For N equations the number of additions and multiplications is $O(N^3)$. The main thrust of so-called quasi-Newton methods is to find an approximation for $\mathbf{J}$, preferably one that reduces the amount of computation to $O(N^2)$. The price paid is that the rate of convergence will be less than quadratic, but not so slow as to negate the reduced amount of computation.

For the case of only two equations above it is convenient to write the equations explicitly, but for three or more it rapidly becomes inconvenient and certainly not to be done on a spreadsheet, even if it is possible to do so.

4.6.5　Example

Apply Newton's method to solve the pair of equations

$$f(x,y) = x^4 + y^4 - 97 = 0$$
$$g(x,y) = x^3 - 3xy + 10 = 0$$

	A	B
1	Example	Newtons method (2 eqns)
2	4.6.5	f(x,y)=x^4+y^4-97=0
3		g(x,y)=x^3-3xy+10=0
4	n	x(n)
5	=0	=1
6	=A5+1	=D5-(-3*D5*D6-4*C5^3*E6)/F6

	C	D
4	y(n)	f(x(n-1),y(n-1))
5	=3.5	
6	=C5-(3*(C5-B5^2)*D6+4*B5^3*E6)/F6	=B5^4+C5^4-97

	E	F
4	g(x(n-1),y(n-1))	J(x(n-1),y(n-1))
5		
6	=B5^3-3*B5*C5+10	=12*((C5-B5^2)*C5^3-B5^4)

	A	B	C
1	Example	Newtons method (2 eqns)	
2	4.6.5	f(x,y)=x^4+y^4-97=0	
3		g(x,y)=x^3-3xy+10=0	
4	n	x(n)	y(n)
5	0	1.0000000	3.5000000
6	1	1.1945752	3.1802286
7	2	1.2955036	3.1179345

	D	E	F
4	f(x(n-1),y(n-1))	g(x(n-1),y(n-1))	J(x(n-1),y(n-1))
5			
6	5.41E+01	5.00E-01	1274.25
7	7.33E+00	3.08E-01	652.26

Screen 4.6.2 *A system of two nonlinear equations solved using Newton's method.*

See Screen 4.6.2. To simplify the formulas the function and Jacobian evaluations are performed in columns D, E, and F, and the iterations (as per equations 4.6.2) are carried out in columns B and C. In addition to the root (2, 3) there are three other roots: (-3.01865, 1.93318), (-0.96801, -3.13116), and (1.31239, 3.11401), to 5 d.p.

Exercise

Experiment with the starting point (x_0, y_0) to find the roots listed above, and to find (roughly) in what region in the (x,y) plane the starting point can lie to give convergence to a given root. This region can be referred to as the "basin of attraction" for that root.

Note: In Example 4.6.5 the basin of attraction for each of the roots (using Newton's method) is quite wide, but this is not always the case. What *can* be relied upon is that, provided the Jacobian matrix is nonsingular in a neighborhood of a root, Newton's method will converge (quadratically), given a starting point "sufficiently near" the root. In many practical applications the finding of such a starting point is the main problem.

Exercises 4.6

(1) Apply Newton's method to find the two roots of the equations

$$f(x, y) = x - 3y + e^x - 4 = 0$$
$$g(x, y) = -2x + y + e^y = 0$$

(2) Derive the formulae (4.6.2) for Newton's method for two equations.

(3) Apply a fixed-point iteration to the equations of Example 4.6.5. You may find that the conditions of Theorem 4.6.2 are not satisfied, but remember that they are not *necessary*, only *sufficient* conditions. One fixed-point form will converge to the root (2, 3), for example, and another will not.

4.6.6 Quasi-Newton Methods

For their enhanced efficiency these methods depend on (i) finding an approximation **A** for **J** that requires less computation, and (ii) the possibility that **A** is constructed in a way that means the computation of the inverse of **A** is $O(N^2)$ rather than $O(N^3)$. Broyden's method is one that satisfies the former requirement, and Dennis & More have shown that the second also applies. As none of these methods are conveniently implemented on a spreadsheet, the interested reader is referred to Burden & Faires [1989] for further details and references.

4.6.7 The Method of Steepest Descent

It was mentioned earlier that the slow but reliable bisection method can be used to get a first approximation to be refined by the faster but sensitive Newton's method. For the multi-variable system of equations there is also a slow but reliable (i.e. globally convergent) method, the method of steepest descent. This method is beyond the realm of convenient work on a spreadsheet. The Solver facility that is built into Excel is based on the method of steepest descent.

§4.7 Multiple Roots

The discussion of multiple roots at the start of this chapter can be expressed a little more formally:

4.7.1 Definition

A solution $x = s$ of equation $f(x) = 0$ is said to have **multiplicity m** if we can write $f(x) = (x - s)^m q(x)$, for $x \neq s$, where $\lim_{x \to s} q(x) \neq 0$.

Newton's method assumes (for quadratic convergence) that the root being sought is a simple root, a fact highlighted by this theorem:

4.7.2 Theorem

$x = s$ is a **simple** root of $f(x) = 0$ if and only if $f(s) = 0$ and $f'(s) \neq 0$.

More generally $x = s$ is a **root of multiplicity** m if and only if the derivatives $f^{(n)}(s) = 0$ for $n = 0, 1, 2, \ldots, m-1$, and $f^{(m)}(s) \neq 0$.

If Newton's method is used to find a root which is not simple then (assuming convergence happens at all) the convergence will *not* be quadratic.

4.7.3 Example

The equation $f(x) = \sin(x^2) = 0$ *has a root of multiplicity 2 at*

$x = 0$, *since*

$$q(x) = \frac{\sin(x^2)}{x^2} \to 1 \text{ as } x \to 0.$$

Use Newton's method to find this root.

In this case the Newton iteration equation 4.4.1 is

$$x_n = x_{n-1} - \frac{\sin(x_{n-1}^2)}{2x_{n-1}\cos(x_{n-1}^2)}, \quad n \geq 1$$

This sequence is tabulated in column B of Screen 4.7.1. Clearly the convergence is not quadratic.

	A	B
1	Example 4.7.3	Newton's method
2		Repeated root at x=0
3		Equation f(x)=sin(x^2)=0
4	n	x(n)
5	=0	=0.2
6	=A5+1	=B5-SIN(B5^2)/(2*B5*COS(B5^2))

	A	B
1	Example 4.7.3	Newton's method
2		Repeated root at x=0
3		Equation f(x)=sin(x^2)=0
4	n	x(n)
5	0	2.00000E-01
6	1	9.99466E-02
7	2	4.99717E-02
8	3	2.49858E-02
9	4	1.24929E-02
10	5	6.24644E-03
11	6	3.12322E-03
12	7	1.56161E-03
13	8	7.80805E-04
14	9	3.90403E-04

Screen 4.7.1 *Newton's method (unmodified) gives only linear convergence to the repeated root at x = 0 of the equation* $f(x) = \sin(x^2) = 0$.

Exercises 4.7.1

(1) Verify that $f(x) = \sin(x^2) = 0$ has a root of multiplicity 2 at $x = 0$ (as defined in Theorem 4.7.2). Repeat for $f(x) = \sin(x) - 1 = 0$ at $x = \frac{\pi}{2}$.

(2) Use the techniques practised in chapter 2 to estimate the (order of) the rate of convergence of the sequence in Example 4.7.3.

4.7.4 Modification of Newton's Method for Multiple Roots.

Consider now the function $p(x) = f(x) / f'(x)$, and assume that $x = s$ is a zero of multiplicity $m \geq 1$ of $f(x)$. From definition 4.7.1 we have

$$f(x) = (x - s)^m q(x),$$

and so

$$p(x) = \frac{(x - s)^m q(x)}{m(x - s)^{m-1} q(x) + (x - s)^m q'(x)}$$

$$= \frac{(x - s) q(x)}{m q(x) + (x - s) q'(x)}$$

Evidently $p(x)$ also has a zero at $x = s$ but it is a simple zero regardless of the value of m. We can apply Newton's method to find s as a zero of $p(x)$ and expect quadratic convergence! The iteration is now $x_n = g(x_{n-1})$ with

$$g(x) = x - \frac{p(x)}{p'(x)} = x - \frac{f(x) f'(x)}{[f'(x)]^2 - f(x) f''(x)}$$

An obvious drawback here is the need for another derivative $f''(x)$. Of some concern too is the possibility of serious rounding errors resulting from the difference of two small numbers in the denominator.

Exercises 4.7.2

(1) Show that $p'(s) = 0$, and derive the last formula above.

(2) Apply this idea to the cases of Exercise 4.7.1(1). Simplify your expression for the second term of $g(x)$, using the trigonometric identities $2 \sin(u) \cos(u) = \sin(2u)$ and $\sin^2(u) + \cos^2(u) = 1$, before going to your spreadsheet.

(3) Apply the iterated Shanks transformation and Steffensen's method to accelerate the sequences of Exercise 4.7.1(1). Compare the speeds obtained with that found in (2) above.

§4.8   Ill-conditioned problems

As already discussed in chapter 3, we say that a problem is ill-conditioned if a small change in the input data causes a large numerical change or a significant qualitative change in the solution(s). An instance of the latter is when the result is a change in the *number* of real roots of an equation.

The variation in the input data could (for example) derive from the uncertainties in experimental measurements, or the limited precision of the computer being used. This latter possibility is demonstrated by the following example, where we use the explicit rounding capability of the spreadsheet. This is provided by the **Round** function, which has the form

Round(expression, r)

As its name suggests, its output value is that of the numerical expression (provided as one of its arguments), after that value has been rounded to r decimal places. It will suffice to illustrate it using simple constants as the expression:

Round(587.6352987, 2)	= 587.64
Round(587.6352987, 3)	= 587.635
Round(587.6352987, -1)	= 590.
Round(583.6352987, -1)	= 580.
Round(587.6352987, -2)	= 600.

Try using **Round** on your spreadsheet in order to be certain of what it does.

4.8.1 Example

Consider the following polynomial which occurs in the study of the motion of a fast moving sub-atomic particle in a metal i.e. the quantum Brownian motion problem:

$$f(x) = (1 + x^{10}) + 10.73327272230(x + x^9)$$
$$+ 51.00379213901(x^2 + x^8)$$
$$+ 141.3652656397(x^3 + x^7) + 253.1594196943(x^4 + x^6)$$
$$+ 306.1293471704 x^5$$

Make two tabulations of this function on the interval [-1.1, -0.7] with 50 sub-intervals, one with the full precision coefficients, and the other where the coefficients are rounded to 6 decimal places. Plot both on the same graph.

See Screen 4.8.1. To avoid very long formulas, the function (in cells B11, B12,...) has been split into a sum of the 6 terms dictated by the definition given above, calculated in D11..I11 and below. The x values are in cells A11, A12,..., and the two sets of coefficients lie in blocks A4..A8 and B4..B8, for the exact and rounded values, respectively. The corresponding graphs are shown in Graph 4.8.1.

It is immediately apparent from the graphs that this small change in the coefficients (of the order of 10^{-6}) has changed the *number* of roots in this range from 4 to 1, and the *value* of the root at (approx) -0.765 to (approximately) -0.775, a change that is 10^4 times larger than the change in the coefficients. The problem is ill-conditioned.

	A	B
1	Example 4.8.1	Ill-conditioned
2	Coefficients	problem
3	exact	6 d.p.
4	10.7332727223	=ROUND(A4,6)
5	51.00379213901	=ROUND(A5,6)
6	141.3652656397	=ROUND(A6,6)
7	253.1594196943	=ROUND(A7,6)
8	306.1293471704	=ROUND(A8,6)

	A
10	x
11	=-1.1
12	=A11+0.008

	B
10	f(x) exact coeffs
11	=D11+A$4*E11+A$5*F11+A$6*G11+A$7*H11+A$8*I11

	C
10	f(x) rounded coeffs
11	=D11+B$4*E11+B$5*F11+B$6*G11+B$7*H11+B$8*I11

	D	E	F
10	$1+x^{10}$	$x+x^9$	x^2+x^8
11	=1+A11^10	=A11+A11^9	=A11^2+A11^8

	G	H	I
10	x^3+x^7	x^4+x^6	x^5
11	=A11^3+A11^7	=A11^4+A11^6	=A11^5

	A	B	C
10	x	f(x) exact coeffs	f(x) rounded coeffs
11	-1.100	5.30E-07	-8.15E-07
12	-1.092	4.77E-07	-8.05E-07

	D	E	F
10	$1+x^{10}$	$x+x^9$	x^2+x^8
11	3.59E+00	-3.46E+00	3.35E+00
12	3.41E+00	-3.30E+00	3.21E+00

	G	H	I
10	x^3+x^7	x^4+x^6	x^5
11	-3.28E+00	3.24E+00	-1.61E+00
12	-3.15E+00	3.12E+00	-1.55E+00

Screen 4.8.1 *Small changes to the coefficients of this polynomial cause 3 of its real zeros to disappear. The problem of finding the zeros of f(x) = 0 is ill-conditioned.*

Clearly it is wise to check the variation of numerically determined roots when small parameter changes are made. Any disproportionate variation would indicate the need for a closer examination of the problem and/or the numerical method used to solve it.

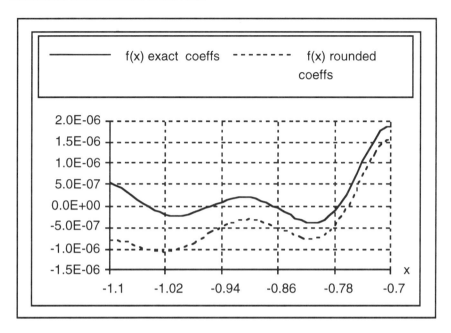

Graph 4.8.1 *The graph of f(x) only crosses the x- axis once (instead of three times) when the coefficients of the polynomial f are changed by as little as 1 part in 10 million.*

§4.9   Cobweb Diagrams

Cobweb diagrams were mentioned in passing earlier in this chapter - see Fig. 4.3.1. Although not directly a part of numerical analysis, an exploration of cobweb diagrams may help the reader to a better understanding of the workings of fixed point iterations. For this reason we include here a spreadsheet that displays the cobweb diagram for any chosen function $g(x)$.

The iteration $x_n = g(x_{n-1})$, $n \geq 1$, with given x_0, generates the sequence of numbers $x_0, x_1 = g(x_0), x_2 = g(x_1), \ldots$ The cobweb diagram traces the evolution of this sequence by plotting a line joining a pair of points that lie alternately on the graphs of $y = g(x)$ and $y = x$. Starting at the initial point x_0 on the x-axis, it plots the following lines:

From $(x_0, 0)$ to (x_0, x_1), where $x_1 = g(x_0)$,

From (x_0, x_1) to (x_1, x_1),

From (x_1, x_1) to (x_1, x_2), where $x_2 = g(x_1)$,

From (x_1, x_2) to (x_2, x_2),

From (x_2, x_2) to (x_2, x_3), where $x_3 = g(x_2)$,

etc.

We first consider the iteration defined by the linear function

$$g(x) = m(x - s) + s$$

which obviously has a fixed point at $x = s$. This case provides the key to understanding the general case because for values of x sufficiently close to a fixed point we can replace $g(x)$ with its tangent line at $x = s$, for which the gradient $m = g'(s)$.

4.9.1 Example

Construct a spreadsheet that generates a cobweb diagram for the linear iteration function $g(x) = m(x-s) + s$, where m and s and the starting point can be varied.

See screen 4.9.1 and Graph 4.9.1. There are two sets of data tabulated. Cells A9..C9 and those below them contain the values of x, the identity function $y = x$, and the iteration function $g(x) = m(x-s) + s$, respectively. Cells G9..H9 and those below them contain the coordinate pairs that produce the cobweb graph, as described above. The IF function switches between the line $y = x$ and the curve $y = g(x)$, controlled by the alternating sign of the number $(-1)^n$ in column F.

This particular iteration has a fixed point at $x = s$. The case depicted has $s = 1$, $m = 0.5$ and $x_0 = 1.9$. The values of m, s, and x_0 are stored in cells C5, C6, and G6, respectively.

It is usual to use arrowed lines when plotting cobweb diagrams (see Fig. 4.2) so that the direction of the evolution of the iteration is clearly indicated. This is not possible on a spreadsheet, which is why the line segment that begins on the horizontal axis is included, It joins $(x_0, 0)$ to (x_0, x_1), where $x_1 = g(x_0)$, and serves to clearly indicate where the sequence begins.

This spreadsheet also allows the user to choose the range $[a, b]$ of values over which the functions $y = g(x)$ and $y = x$ are plotted, and this range is subdivided into 50 steps, a choice that the user can easily change if he or she wishes.

The range of x values $[a, b]$ plotted and the step-size used may be varied by changing the relevant entries in cells D1, D2, and D3, amending (if necessary) the number of FILLed rows in columns A, B and C, and then re-creating the plot.

	A	B	C	D
1	Cobweb Diagram		a =	=0
2	for fixed point		b =	=2
3	iteration		step-size =	=(D2-D1)/50
4	x(n+1)=g(x(n))			
5	g(x)=m(x-s)+s	m =	=0.5	
6		s =	=1	
7	Data for plots y=x	and	y=g(x)	
8	x	y=x	y=g(x)	
9	=D1	=A9	=C$5*(A9-C$6)+C$6	
10	=A9+D$3	=A10	=C$5*(A10-C$6)+C$6	

	E	F	G	H
3		(50 steps		
4		plotted)		
5				
6			x(0) =	=1.9
7			Data for cobweb plot	
8	n	(-1)^n	x(cobweb)	y(cobweb)
9	=0	=(-1)^E9	=H6	=0
10	=E9+1	=(-1)^E10	=IF(F10<0,G9,H9)	=IF(F10<0,C$5*(G10-C$6)+C$6,G10)

	A	B	C	D
1	Cobweb Diagram		a =	0
2	for fixed point		b =	2
3	iteration		step-size =	0.04
4	x(n+1)=g(x(n))			
5	g(x)=m(x-s)+s	m =	0.5	
6		s =	1	
7	Data for plots y=x and y=g(x)			
8	x	y=x	y=g(x)	
9	0	0	0.5	
10	0.04	0.04	0.52	
11	0.08	0.08	0.54	

	E	F	G	H
6			x(0) =	1.9
7			Data for cobweb plot	
8	n	(-1)^n	x(cobweb)	y(cobweb)
9	0	1	1.9	0
10	1	-1	1.9	1.45
11	2	1	1.45	1.45

Screen 4.9.1 *This spreadsheet generates a cobweb diagram for the fixed point iteration defined by the linear function $g(x) = m(x - s) + s$.*

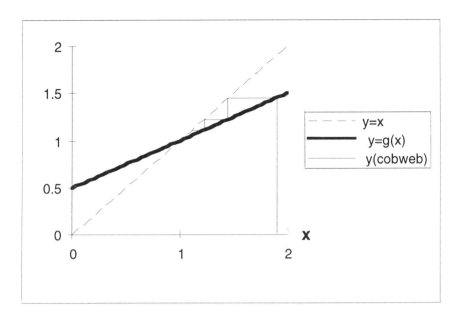

Graph 4.9.1 *The cobweb diagram that is produced by the spreadsheet of Screen 4.9.1, when s = 1 and m = 0.5.*

Exercises 4.9

(1) Vary the starting value x_0 and try other values of m, for instance: -0.5, 1.5, -1.5, and -1.0.

(2) Modify the spreadsheet so that it displays the cobweb diagram for a different iteration, for example $g(x) = \sqrt{x}$.

(3) Use the cobweb diagram spreadsheet to study the iterations listed in Exercise 4.3.1(2).

5 Numerical Integration

Many definite integrals cannot be exactly evaluated by hand. A simple yet important example is the integral which defines the "error" function $erf(x)$:

$$erf(x) = \frac{2}{\sqrt{\pi}} \int_0^x e^{-t^2} dt$$

In this chapter we consider some methods for finding an estimate for the integral

$$\int_a^b f(x) dx \ .$$

The integrand $f(x)$ is assumed to be given explicitly or to be otherwise available numerically, and we will assume that $a < b$. The methods considered will include error bounds for the approximations they provide.

The idea behind the methods to be studied is the replacement of function f with a sequence of simpler functions on N adjacent sub-intervals of the interval $[a,b]$. The original integral is then approximated by a sum of simpler integrals, leading to a rule that gives an estimate of the integral in terms of the value of the integrand at particular points in each sub-interval.

§5.1 Approximation of Integrals

If the interval $[a,b]$ is subdivided into N equal parts, each of length $h = (b-a)/N$, then

$$[a,b] = [a, a+h] \cup [a+h, a+2h] \cup ... \cup [a+(N-1)h, b]$$

and

$$\int_a^b f(x)dx = \int_a^{a+h} f(x)dx + \int_{a+h}^{a+2h} f(x)dx + ... + \int_{a+(N-1)h}^b f(x)dx.$$

Next, we make the approximation

$$\int_a^b f(x)dx \approx I_N = \int_a^{a+h} f_1(x)dx + \int_{a+h}^{a+2h} f_2(x)dx + ... + \int_{a+(N-1)h}^b f_N(x)dx$$

where $f_n(x)$ is the function used to approximate $f(x)$ on the nth sub-interval. It is assumed that the anti-derivative of the function $f_n(x)$ is easily available "by hand". The methods to be considered here employ either a linear or a quadratic approximating function, both of which are trivial to integrate.

Finally, as the length h of the sub-intervals tends to zero (and $N \to \infty$) the exact value of the integral is approached:

$$\int_a^b f(x)dx = \underset{N \to \infty}{\text{limit}}\, I_N,$$

assuming this limit to exist. In practice N is increased only as far as is needed to reach the required accuracy.

A simple geometric motivation will be given for the derivation of the numerical integration formulas. More sophisticated methods (not discussed here - see Burden & Faires [1989], Schwarz [1989], for instance) are needed to find the error terms quoted later on.

§5.2 The Midpoint and Trapezoidal Rules

5.2.1 The Midpoint Rule

On each sub-interval the function is replaced by a constant, in this case the value of the function at the midpoint of the sub-interval:

$$f_n(x) = f(a + (n - \tfrac{1}{2})h)$$

The crudest approximation is when $N = 1$ i.e. $h = b - a$, illustrated in Fig. 5.2.1. Here, the midpoint lies at $c = (a + b)/2 = a + \tfrac{1}{2}h$, and

$$\int_a^b f(x)dx \approx hf(c) = hf(c) = hf(a + \tfrac{1}{2}h) = \text{ area of shaded rectangle}$$

With **2 sub-intervals**, $h = (b - a)/2$, shown in Fig. 5.2.2,

$$\int_a^b f(x)dx \approx h[f(a + \tfrac{1}{2}h) + f(a + \tfrac{3}{2}h)]$$

$$= \text{sum of the areas of the two shaded rectangles}$$

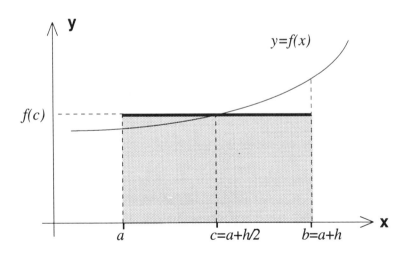

Fig. 5.2.1 *Estimating an integral using the midpoint rule with 1 sub-interval.*

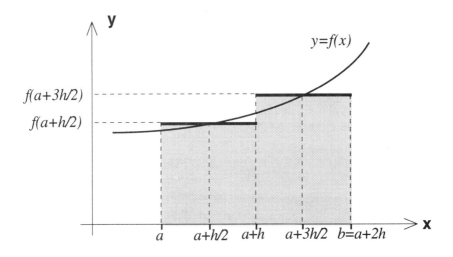

Fig. 5.2.2 *Estimating an integral using the midpoint rule with 2 sub-intervals.*

With **3 sub-intervals**, $h = (b - a)/3$, and

$$\int_a^b f(x)dx \approx h[f(a + \tfrac{1}{2}h) + f(a + \tfrac{3}{2}h) + f(a + \tfrac{5}{2}h)]$$

$$= \text{sum of the areas of three rectangles}$$

With N **sub-intervals**, $h = (b - a)/N$ and

Midpoint Rule

$$\int_a^b f(x)dx \approx h[f(a + \tfrac{1}{2}h) + f(a + \tfrac{3}{2}h) + ... + f(a + \tfrac{1}{2}(2N - 1)h)] \quad (5.2.1)$$

$$= h\sum_{m=1}^{N} f(a + (m - \tfrac{1}{2})h)$$

Note: Two similar rules can be developed using the value of the function at either the left-hand end or the right-hand end of each sub-interval instead of the midpoint value.

Exercise

Develop formulae similar to eq. 5.2.1 above for these alternative methods.

5.2.2 The Trapezoidal Rule

Here, on each sub-interval, the function f is replaced by a straight line joining the endpoints of the sub-interval (on the graph of f). The area under the curve on each sub-interval is now approximated by the area of the resulting trapezium.

With $N = 1$ i.e. $h = b - a$, we have (see Fig. 5.2.3):

$$\int_a^b f(x)dx \approx \frac{1}{2}(b-a)[f(a) + f(b)] = h[\tfrac{1}{2}f(a) + \tfrac{1}{2}f(b)]$$

$=$ area of shaded trapezium

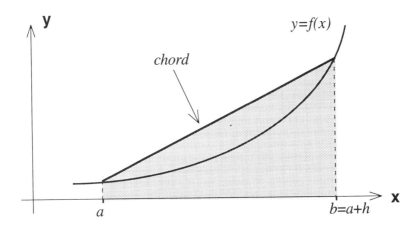

Fig. 5.2.3 *Estimating an integral using the trapezoidal rule with 1 sub-interval.*

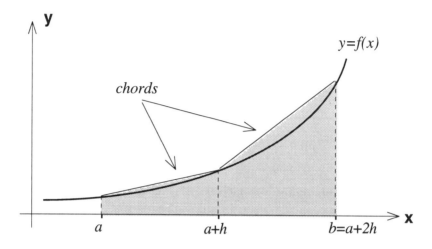

Fig. 5.2.4 *Estimating an integral using the trapezoidal rule with 2 sub-intervals.*

With two sub-intervals, $N = 2$, $h = (b - a)/2$, and (see Fig. 5.2.4):

$$\int_a^b f(x)dx \approx \text{area(1st trapezium)} + \text{area(2nd trapezium)}$$

$$= h[\tfrac{1}{2} f(a) + \tfrac{1}{2} f(a+h)] + h[\tfrac{1}{2} f(a+h) + \tfrac{1}{2} f(b)]$$

$$= h[\tfrac{1}{2} f(a) + f(a+h) + \tfrac{1}{2} f(b)]$$

With three sub-intervals, $N = 3$, $h = (b - a)/3$, and

$$\int_a^b f(x)dx \approx \text{sum of the areas of three trapeziums}$$

$$= h[\tfrac{1}{2} f(a) + f(a+h) + f(a+2h) + \tfrac{1}{2} f(b)]$$

For N sub-intervals, $h = (b - a)/N$, and

Trapezoidal Rule

$$\int_a^b f(x)dx \approx h[\tfrac{1}{2} f(a) + \sum_{m=1}^{N-1} f(a+mh) + \tfrac{1}{2} f(b)] \qquad (5.2.2)$$

Although the purpose of this chapter is to explore methods for evaluating integrals that can only be found numerically, it is reassuring to first try them out with such examples as the following, where the result is known beforehand.

5.2.3   Example

Apply the midpoint and trapezoidal methods to the integral

$$I_1 = \int_0^1 x\exp(-x^2)dx = \tfrac{1}{2}(1-\tfrac{1}{e}) \approx 0.3160603$$

In this case, $a = 0, b = 1$, $f(x) = x\exp(-x^2)$ and $h = 1/N$.

The spreadsheets shown in Screens 5.2.1 and 5.2.2 have been designed to make easy any variation of the parameters a, b, and N, and a change of integrand $f(x)$. In both spreadsheets, a, b, and N are stored in cells D1, D2, and D3, respectively, and h is calculated in cell D4. The integrand is tabulated (as needed) in column B, alongside the required x values in column A, starting at row 7.

For the **midpoint rule** (equation 5.2.1, Screen 5.2.1) we need to form the sum
$$S_{midpt} = f(a+\tfrac{1}{2}h) + f(a+\tfrac{3}{2}h) + ... + f(a+\tfrac{1}{2}(2N-1)h).$$

Note that each term in this sum has the same "weight" (=1). The sum is calculated in cell C7 and those below it, and the estimated integral is obtained in column D when the sum is multiplied by h. The IF function in column D serves to ensure the display of a non-zero result only when all N terms have been summed. The example displayed has $N = 5$. For larger N, simply change the N value in cell D3 and FILL down enough rows (i.e. down to where $n=N$) to reveal the result. The results obtained with $N = 5, N = 10$ and $N = 20$ are shown below.

h	I_1 (5 d.p.)	|error| (2 s.f.)
0.2	0.31836	0.0023
0.1	0.31663	0.00057
0.05	0.31620	0.00014

An alternative to the running sum in column C is to use the spreadsheet SUM function, replacing the formula shown in cell C7 with =SUM(b$7..B7), and then FILL down to the row with $n = N$.

	A	B
1	Example	Numerical
2	5.2.3a	Integration
3		
4	Midpoint	rule
5		f(x)=
6	x	xexp(-x^2)
7	=D1+D4/2	=A7*EXP(-(A7^2))
8	=A7+D$4	=A8*EXP(-(A8^2))

	C	D	E
1	a =	=0	
2	b =	=1	
3	N =	=5	
4	h =	=(D2-D1)/D3	
5		Integral	
6	Sum	(in row with n=N)	n
7	=B7	=IF(E7=D$3,1,0)*D$4*C7	=1
8	=C7+B8	=IF(E8=D$3,1,0)*D$4*C8	=E7+1

	A	B	C	D	E
1	Example	Numerical	a =	0	
2	5.2.3a	Integration	b =	1	
3			N =	5	
4	Midpoint	rule	h =	0.2	
5		f(x)=		Integral	
6	x	xexp(-x^2)	Sum	(in row with n=N)	n
7	0.1	0.099005	0.099005	0.000000	1
8	0.3	0.274179	0.373184	0.000000	2
9	0.5	0.389400	0.762585	0.000000	3
10	0.7	0.428838	1.191423	0.000000	4
11	0.9	0.400372	1.591795	0.318359	5

Screen 5.2.1 *Numerical integration using the midpoint rule.*

	A	B	C
1	Example	Numerical	a =
2	5.2.3b	Integration	b =
3			N =
4	Trapezoidal	rule	h =
5		f(x)=	
6	x	xexp(-x^2)	n
7	=D1	=A7*EXP(-(A7^2))	=0
8	=A7+D$4	=A8*EXP(-(A8^2))	=C7+1

	D	E	F
1	=0		
2	=1		
3	10		
4	=(D2-D1)/D3		Integral
5		weighted	(in row
6	weights	sum	with n=N)
7	=IF(OR(C7=0,C7=D$3),0.5,1)	=D7*B7	=(D$4*E7)*IF(C7=D$3,1,0)
8	=IF(OR(C8=0,C8=D$3),0.5,1)	=E7+D8*B8	=(D$4*E8)*IF(C8=D$3,1,0)

	A	B	C	D	E	F
1	Example	Numerical	a =	0		
2	5.2.3b	Integration	b =	1		
3			N =	5		
4	Trapezoidal	rule	h =	0.2		Integral
5		f(x)=			weighted	(in row
6	x	xexp(-x^2)	n	weights	sum	with n=N)
7	0	0.000000	0	0.5	0.000000	0.000000
8	0.2	0.192158	1	1	0.192158	0.000000
9	0.4	0.340858	2	1	0.533015	0.000000
10	0.6	0.418606	3	1	0.951621	0.000000
11	0.8	0.421834	4	1	1.373455	0.000000
12	1	0.367879	5	0.5	1.557395	0.311479

Screen 5.2.2 Numerical integration using the trapezoidal rule.

In the case of the **trapezoidal rule** (equation 5.2.2, Screen 5.2.2), we require the sum

$$S_{trap} = \tfrac{1}{2} f(a) + f(a+h) + f(a+2h) + \ldots + f(a+(N-1)h) + \tfrac{1}{2} f(b)$$

Here, the first ($n=0$) and last ($n=N$) terms have weight $= \tfrac{1}{2}$ and the "interior" ($n=1, 2,\ldots,N-1$) terms have weight =1. The formulas in cells D7 and below give the correct weights, as described below.

The IF function returns the value 0.5 if either $n = 0$ or $n = N$, and the value 1 otherwise.

The correctly weighted running sum is assembled in column E, and the estimated integral is shown in column F, in the row where $n = N$.

Extension to larger values of N than the case $N = 5$ shown is done the same way as for the midpoint rule spreadsheet. Results for $N = 5$, $N = 10$ and $N = 20$ are shown below.

h	I_1 (5 d.p.)	|error| (2 s.f.)
0.2	0.31148	0.0046
0.1	0.31492	0.0011
0.05	0.31577	0.00029

Note: It appears that for both midpoint and trapezoidal rules the error is reduced by factor of about a quarter each time h is halved. The reason for this will be seen in the later discussion of error bounds.

Exercises 5.2

Apply both rules to estimate the integrals

$$(1) \int_1^2 \frac{dx}{x} \quad (2) \int_2^4 x\sqrt{1+x^2}\,dx \quad (3) \int_0^{2\pi} x\sin(x)\,dx \quad (4)\ \mathrm{erf}(1.5)$$

Begin with $N = 5$ and then extend your spreadsheet in each case for $N = 10$ and $N = 20$. For the first three integrals compare the results with the exact value and find the error. The exact values are $\ln(2)$, $\frac{1}{3}(17^{\frac{3}{2}} - 5^{\frac{3}{2}})$ and -2π, respectively, for the first three integrals.

§5.3 Simpson's Rule

Here, $f(x)$ is replaced (on each sub-interval) by a quadratic function $Q(x)$ which is chosen so that $f(x)$ and $Q(x)$ have the same values at the end points and at the midpoint of the sub-interval.

We begin by considering the case $N=1$, i.e. $h=b-a$. The desired result is more easily found if we translate the x-coordinate so that the interval is symmetric about the new origin. Let $x=s+c$ where $c=\frac{1}{2}(a+b)$. It follows that

$$I = \int_a^b f(x)dx = \int_{-\frac{1}{2}h}^{\frac{1}{2}h} f(c+s)ds$$

Now let $Q(s)=As^2+Bs+C$, where A, B and C will be chosen so that $Q(s)$ and $f(c+s)$ will have the same value at the three points $s=-\frac{1}{2}h$, $s=0$ and $s=\frac{1}{2}h$, corresponding to $x=a$, $x=c=\frac{1}{2}(a+b)$ and $x=b$ respectively. Hence we have

$$Q(-\tfrac{1}{2}h)=\tfrac{1}{4}Ah^2-\tfrac{1}{2}Bh+C=f(c-\tfrac{1}{2}h)=f(a)$$
$$Q(0)=C=f(c)$$
$$Q(\tfrac{1}{2}h)=\tfrac{1}{4}Ah^2+\tfrac{1}{2}Bh+C=f(c+\tfrac{1}{2}h)=f(b)$$

Next,

$$I \approx \int_{-\frac{1}{2}h}^{\frac{1}{2}h} Q(s)ds = \tfrac{1}{12}Ah^3+Ch$$

But

$$A=\frac{2}{h^2}[f(a)-2f(c)+f(b)] \text{ and } C=f(c)$$

so that

$$I \approx \tfrac{1}{6}h[f(a)+4f(c)+f(b)]$$
$$= \tfrac{1}{6}h[f(a)+4f(a+\tfrac{1}{2}h)+f(a+h)]$$

With two sub-intervals, $N=2$, $h=(b-a)/2$, we apply the result above to each of the sub-intervals $[a,a+h]$ and $[a+h,b]$ in turn, to find

$$\int_a^b f(x)dx = \int_a^{a+h} f(x)dx + \int_{a+h}^b f(x)dx$$

$$\approx \tfrac{1}{6}h[f(a) + 4f(a+\tfrac{1}{2}h) + f(a+h)]$$

$$+ \tfrac{1}{6}h[f(a+h) + 4f(a+\tfrac{3}{2}h) + f(b)]$$

$$= \tfrac{1}{6}h[f(a) + 4\{f(a+\tfrac{1}{2}h) + f(a+\tfrac{3}{2}h)\} + 2f(a+h) + f(b)]$$

With three sub-intervals, $N = 3$, $h = (b-a)/3$, we get

$$\int_a^b f(x)dx \approx \tfrac{1}{6}h[f(a) + 4\{f(a+\tfrac{1}{2}h) + f(a+3(\tfrac{1}{2}h)) + f(a+5(\tfrac{1}{2}h))\}$$

$$+ 2\{f(a+2(\tfrac{1}{2}h)) + f(a+4(\tfrac{1}{2}h))\} + f(b)]$$

With N sub-intervals, $h = (b-a)/N$, and we deduce

Simpson's Rule

$$\int_a^b f(x)dx \approx \tfrac{1}{6}h[f(a) + 4\sum_{m=1}^{N} f(a+\tfrac{1}{2}(2m-1)h)$$

$$+ 2\sum_{m=1}^{N-1} f(a+mh) + f(b)] \qquad (5.3.1)$$

5.3.1   Example

Apply Simpson's rule (with $N = 2$) to the integral of Example 5.2.3,

$$I_1 = \int_0^1 x\exp(-x^2)dx = \tfrac{1}{2}(1 - \tfrac{1}{e}) \approx 0.3160603$$

It is evident from equation 5.3.1 that we need to form the sum

$$S_{Simpson} = f(a) + 4 \sum_{m=1}^{N} f(a + \tfrac{1}{2}(2m-1)h) + 2 \sum_{m=1}^{N-1} f(a+mh) + f(b)$$

$$= \sum_{n=1}^{2N} w_n f(a + \tfrac{1}{2}nh), \text{ say.}$$

The weights w_n are given by

$$w_n = \begin{cases} 1 & n = 0 \text{ or } n = 2N \\ 4 & n \text{ odd} \\ 2 & n \text{ even}, \ n \neq 0, \ n \neq 2N \end{cases}$$

For example, when $N = 1$ this sum is $f(a) + 4f(a + \tfrac{1}{2}h) + f(b)$, and the weights are 1, 4, and 1. For $N = 2$ the sum is $f(a) + 4f(a + \tfrac{1}{2}h) + 2f(a+h) + 4f(a + \tfrac{3}{2}h) + f(b)$, with weights 1, 4, 2, 4, and 1.

We need a tabulation with step-size $\tfrac{1}{2}h$. A suitable spreadsheet is shown in Screen 5.3.1, for the case $N = 2$. The formulae in columns C and D are contrived to provide the weights (as described below), the weighted sum is calculated in column E, and the estimated integral is revealed in column F, in the row where $n = 2N$.

The weights are calculated as follows: Divide either 2 or 4 (when n is even or odd, respectively) by either 2 or 1 (when $n = 0$ or $n = 2N$ on the one hand, or neither, on the other). This gives $2/2 = 1$ when $n = 0$ or $n = 2N$, $4/2 = 2$ when n is even (other than $n = 0$ or $n = 2N$), and $4/1 = 4$ when n is odd.

To extend the spreadsheet to larger N, enter the new N in cell D3, and FILL columns A to F down until the row with $n = 2N$ is reached. To integrate a different function, enter its spreadsheet formula in cell B7 and FILL down from there.

The result $I_1 \approx 0.3162868$, for which $|error| \approx 0.0002$, which is more accurate than found with the midpoint and trapezoidal rules with $N = 10$.

With $N = 4$ Simpson's rule gives $|error| \approx 0.000013$. It would seem that for a given number of sub-intervals Simpson's rule can be *much* more accurate than both the midpoint and trapezoidal rules.

	A	B
1	Example	Numerical
2	5.3.1	Integration
3	Simpson's	rule
4		
5		f(x) =
6	x	xexp(-x^2)
7	=0	=A7*EXP(-(A7^2))
8	=A7+D$4/2	=A8*EXP(-(A8^2))

	C	D
1	a =	=0
2	b =	=1
3	N =	2
4	h =	=(D2-D1)/D3
5		
6	n	weights
7	=0	=IF((-1)^C7>0,2,4)/IF(OR(C7=0,C7=2*D$3),2,1)
8	=C7+1	=IF((-1)^C8>0,2,4)/IF(OR(C8=0,C8=2*D$3),2,1)

	E	F
5	weighted	
6	sum	Integral
7	=D7*B7	=IF(C7=2*D$3,D$4*E7/6,"*")
8	=E7+D8*B8	=IF(C8=2*D$3,D$4*E8/6,"*")

	A	B	C	D	E	F
1	Example	Numerical	a =	0		
2	5.3.1	Integration	b =	1		
3	Simpson's	rule	N =	2		
4			h =	0.5		
5		f(x) =			weighted	
6	x	xexp(-x^2)	n	weights	sum	Integral
7	0	0	0	1	0.00000	*
8	0.25	0.23485	1	4	0.93941	*
9	0.5	0.38940	2	2	1.71821	*
10	0.75	0.42734	3	4	3.42756	*
11	1	0.36788	4	1	3.79544	0.3162868

Screen 5.3.1 Numerical integration using Simpson's rule.

Exercises 5.3

Apply Simpson's rule to the integrals listed in Exercises 5.2. How small can N be and still reach at least the same accuracy achieved with $N = 20$ when using the midpoint rule?

§5.4 The Error Term, R_N

The formulae for the 3 methods can be expressed in the following form:

5.4.1 midpoint rule

$$\int_a^b f(x)dx = h\sum_{m=1}^{N} f(a + \tfrac{1}{2}(2m-1)h) + R_N$$

where

$$R_N = \frac{h^3 N f''(\xi_N)}{24} = \frac{h^2(b-a)f''(\xi_N)}{24}, \text{ for some } \xi_N \in [a,b] \qquad (5.4.1)$$

5.4.2 trapezoidal rule

$$\int_a^b f(x)dx = \tfrac{1}{2}h[f(a) + 2\sum_{m=1}^{N-1} f(a+(m-1)h) + f(b)] + R_N$$

where

$$R_N = -\frac{h^3 N f''(\xi_N)}{12} = -\frac{h^2(b-a)f''(\xi_N)}{12}, \text{ for some } \xi_N \in [a,b] \qquad (5.4.2)$$

5.4.3 Simpson's rule

$$\int_a^b f(x)dx = \tfrac{1}{6}h[f(a) + 4\sum_{m=1}^{N} f(a + \tfrac{1}{2}(2m-1)h)$$

$$+ 2\sum_{m=1}^{N-1} f(a+mh) + f(b)] + R_N$$

where

$$R_N = -\tfrac{1}{90}N(\tfrac{h}{2})^5 f^{(4)}(\xi_N)$$

$$= -\tfrac{1}{180}(b-a)(\tfrac{h}{2})^4 f^{(4)}(\xi_N), \text{ for some } \xi_N \in [a,b] \qquad (5.4.3)$$

In each case, for a given N, i.e. given h, the maximum size of the error is given by the maximum value of $|R_N|$ on the interval $[a,b]$. An upper bound for $|R_N|$ will be provided by knowledge of an upper bound for $|f''|$ (in the midpoint and trapezoidal cases) or for $|f^{(4)}|$ (for Simpson's rule) on $[a,b]$.

Note: The midpoint and trapezoidal rules give *exact* results for polynomials of degree less than or equal to 1, and Simpson's rule is *exact* for polynomials of degree less than or equal to 3.

5.4.4 O Notation

A function $g(h)$ of h is said to be $O(h^p)$ (i.e. $g(h)$ is of order h^p) if (i) $g(h)/h^p$ is bounded as $h \to 0$, but (ii) $g(h)/h^q$ is unbounded (as $h \to 0$) if $q > p$.

For example, consider $g(h) = 5h^3 + h^4$. Here,

$$\lim_{h\to 0} \frac{g(h)}{h^{3+t}} = \lim_{h\to 0}(\frac{5}{h^t} + h^{1-t}) = 5$$

if $t = 0$, and the limit does not exist for $t > 0$. Hence $g(h)$ is $O(h^3)$. To put it another way, $g(h) \approx 5h^3$ if h is very small.

We can now make the following observations:

$|R_N|$ is $O(h^2)$ for the midpoint and trapezoidal rules, and $O(h^4)$ for Simpson's rule. Thus, for example, if h is halved then the upper bound for $|R_N|$ is approximately quartered in the case of the first two rules, and approximately reduced by a factor of $\frac{1}{16}$ for the last rule. The actual error will be reduced by at least as much.

5.4.5 Example

(a) Find error bounds for the integrations discussed in the preceding two examples and compare them with the actual errors found there, and (b) find the minimum number of sub-intervals needed for the error to have magnitude less than 10^{-10} when Simpson's rule is used.

(a) For the integrand of the preceding examples, $f(x) = x\exp(-x^2)$ we have

$$f''(x) = -2x(3-2x^2)\exp(-x^2)$$

and

$$f^{(4)}(x) = 4x(15-20x^2+4x^4)\exp(-x^2).$$

When these derivatives are plotted for $x \in [0,1]$ (see Graph 5.4.1) it is clear that on this interval we can certainly claim that $|f''(x)| < 2$ and $|f^{(4)}(x)| < 17$. These inequalities allow us to find *upper bounds* for the magnitude of the errors in the estimated integrals considered earlier. Thus for the **midpoint rule** (equation 5.4.1 with $N = 5, h = 0.2 = \frac{1}{5}$) we have

$$|R_N| < \frac{1}{5^3} \times \frac{5 \times 2}{24} = \frac{1}{3} \times 10^{-2} \approx 0.0033$$

which is larger than the actual error (≈ 0.002).

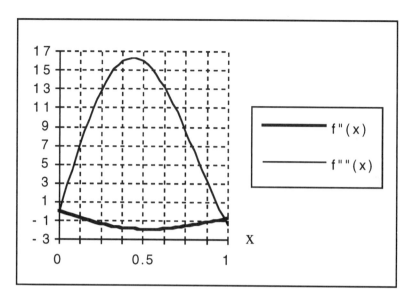

Graph 5.4.1 *Plots of the 2nd and 4th derivatives of $f(x) = x\exp(-x^2)$ on $[0,1]$.*

For the **trapezoidal** rule (equation 5.4.2),

$$|R_N| < \frac{2}{3} \times 10^{-2} \approx 0.0067,$$

which is larger than the actual error (≈ 0.0046).

For **Simpson's** rule (equation 5.4.3 with $N = 2$, $h = \frac{1}{2}$) we get

$$|R_N| < \frac{1}{90} \times 2 \times \frac{1}{4^5} \times 17 = \frac{17}{45 \times 1024} \approx 0.00037$$

i.e. greater than the actual error (≈ 0.0002).

(b) If we want to find the minimum number of sub-intervals needed to reach a given accuracy, the formulas for $|R_N|$ will suffice, provided upper bounds for the size of the relevant derivatives can be found.

Suppose, for example, that we want to know the minimum N for which Simpson's rule will estimate the integral of Example 5.3.1 with absolute error of at most 10^{-10}. Now in this case $h = 1/N$ and

$$|R_N| = \frac{1}{90}\frac{Nh^5}{2^5}|f^{(4)}(x)|$$

$$= \frac{1}{90}\frac{N}{(2N)^5}|f^{(4)}(x)| < \frac{1}{90}\frac{N}{(2N)^5}\times 17$$

Hence $|R_N| \leq 10^{-10}$ if $N^4 > (17\times 10^{10})/(90\times 32)$ i.e. $N \geq 88$. The required accuracy *may* be reached with a value of N less than 88, using 88 will guarantee it.

5.4.6   Example

Use Simpson's rule to estimate the integral

$$I_2 = \int_1^5 \ln(x)dx = 5\ln(5) - 4 \approx 4.047190$$

with error not exceeding 10^{-4}.

Calculations similar to those above indicate $|\text{error}| \leq 10^{-4}$ will be guaranteed if $N \geq 13$. Screen 5.4.1 shows the spreadsheet with $N = 12$ in which case the actual error is less than 10^{-5}. The weights formulae (columns C and D and columns A, E and F are the same as in Screen 5.3.1, and only the formulas in column B (defining the integrand function) need to be changed, as shown.

	B
6	f(x)=ln(x)
7	=LN(A7)
8	=LN(A8)

	A	B	C	D	E	F
6	x	f(x)=ln(x)	n	weights	sum	Integral
31	5.000000	1.609438	24	1	72.849	4.047181

Screen 5.4.1 *Another example of numerical integration with Simpson's rule.*

Exercises 5.4

(1) Verify the calculation of minimum N mentioned in Example 5.4.6.

(2) Find the minimum number of sub-intervals needed if the midpoint rule is used instead of Simpson's rule in Example 5.4.6.

(3) Set up a spreadsheet that will enable calculation of the function $erf(x)$ using Simpson's rule with $N = 10$ and various values of x, up to (say) $x = 5$. Find the error bounds and try other values of N. Compare your values with those listed in tables of mathematical functions, for instance Abramowitz & Stegun[1965].

Note: Increasing the value of N means also increasing the number of function evaluations and there will be more terms in the sums that constitute the methods considered. This has the potential that the rounding error might increase as N increases and h decreases. In fact these methods are **stable** in this sense: the rounding error is *independent* of h (see Burden & Faires [1989]).

§5.5 Integration over an Infinite Interval

Special methods have been developed for integrations over an infinite interval such as

$$\int_a^\infty f(x)dx \quad and \quad \int_{-\infty}^\infty f(x)dx$$

The first of these integrals (for $a > 0$), if it exists, can often be converted to an integral with a finite range of integration by using the change of variable $t = x^{-1}$, in which case

$$\int_a^\infty f(x)dx = \int_0^{a^{-1}} t^{-2} f(t^{-1})dt \qquad (5.5.1)$$

This method is one of a class called **Transformation methods,** developed specifically to deal with improper integrals like those above and integrals where the integrand is unbounded in the range of integration (and the integral exists). See Schwarz[1989] for further details and references.

5.5.1 Example

Apply the transformation (5.5.1) to find

$$\int_y^\infty x\exp(-x^2)dx = \int_0^{y^{-1}} t^{-3}\exp(-t^{-2})dt = \int_0^{y^{-1}} g(t)dt$$

using Simpson's rule.

The spreadsheet of Example 5.3.1 is readily adapted to this purpose: just change the integrand function in column B and add a cell (e.g. F3) for the value of y. Only these changes are given in Screen 5.5.1. To avoid a "divide by zero" error condition we need to explicitly put $g(0) = 0$ in cell B7, valid since $\lim_{t\to 0} t^{-3}\exp(-t^{-2}) = 0$.

Further, we must accomodate the fact that the spreadsheet may give an error condition when it attempts to evaluate $g(t)$ for small values of t Both these needs are satisfied by assigning $g(t)$ the value zero for $t < 0.05$ say, as shown in Screen 5.5.1. As you can see there when you set up the spreadsheet yourself, $g(0.05)$ is about 10^{-170}, so this seems a safe choice given that (for $y = 5$) elsewhere $g(t)$ is about 10^{-12}. Some results are shown below.

N	Integral for y = 2 (6 s.f.)
10	$9.15756.10^{-3}$
15	$9.15777.10^{-3}$
20	$9.15780.10^{-3}$

N	Integral for $y = 5$ (6 s.f.)
40	$6.94782 \cdot 10^{-12}$
45	$6.94639 \cdot 10^{-12}$
50	$6.94556 \cdot 10^{-12}$

	A	B
1	Example	Numerical
2	5.5.1	integration
3	Simpson's	rule
4		
5		g(t)=
6	t	t^(-3)exp(-t^(-2))
7	=D1	=IF(A7<0.05,0,EXP(-1/(A7^2))/(A7^3))
8	=A7+D$4/2	=IF(A8<0.05,0,EXP(-1/(A8^2))/(A8^3))

	C	D	E	F
1	a =	0		
2	b =	=1/F3		
3	N =	3	y =	=5
4	h =	=(D2-D1)/D3		

	A	B	C	D	E	F
1	Example	Numerical	a =	0		
2	5.5.1	integration	b =	0.2		
3	Simpson's rule		N =	40	y =	5
4			h =	0.005		
5		g(t)=			weighted	
6	t	t^(-3)exp(-t^(-2))	n	weights	sum	Integral
86	0.1975	9.5351E-10	79	4	0.0000	*
87	0.2000	1.7360E-09	80	1	0.0000	6.94782E-12

Screen 5.5.1

We consider next an alternative approach to an integral over a semi-infinite interval, which we can split up as follows:

$$\int_0^\infty f(x)dx = \int_0^Y f(x)dx + \int_Y^\infty f(x)dx$$

It is quite often the case that a major part of the integral comes from a finite part [0,Y] of the interval, with Y chosen large enough for the remaining part (the second integral) to be (relatively) very small. It is

172

often possible to get an analytic estimate of an upper bound for that remainder.

5.5.2 Example

We consider the integral

$$I = \int_0^\infty x \exp(-x^2)dx = \int_0^Y x \exp(-x^2)dx + \int_Y^\infty x \exp(-x^2)dx$$

for which the exact value is easily found:

$$I = \lim_{a \to \infty} \int_0^a x e^{-x^2} dx = \lim_{a \to \infty} \frac{1}{2}(1 - \exp(-a^2)) = \frac{1}{2}$$

Let's look at the second term $K(Y)$, where

$$K(Y) = \int_Y^\infty x e^{-x^2} dx = \lim_{b \to \infty} \frac{1}{2}(\exp(-Y^2) - \exp(-b^2)) = \frac{1}{2}\exp(-Y^2).$$

Taking $Y = 5$ gives $K(5) = e^{-25}/2 \approx 7 \times 10^{-12}$, which agrees with the value found for this integral in Example 5.5.1.

Clearly, replacing I with the integral from 0 to 5 will give a very good estimate of I. The major source of error is to be found in the numerical method we use to find the latter integral.

Exercises 5.5

(1) Apply equation 5.5.1 to the integral

$$\int_1^\infty \sqrt{x} e^{-x} dx.$$

Estimate the resulting integral using the trapezoidal rule with $N = 20$ and Simpson's rule with $N = 5$. In both cases vary N in order to observe the resulting changes and find the minimum N needed to get an accuracy of 0.1%.

(2) Consider

$$\int_0^\infty \frac{dx}{1+x^2} = \int_0^1 \frac{dx}{1+x^2} + \int_1^\infty \frac{dx}{1+x^2} = \frac{\pi}{2}$$

Estimate both integrals on the right-hand side using the midpoint rule, applying equation 5.5.1 to the second integral, and compare the result with the exact value.

(3) Adapt the spreadsheets of Examples 5.2.3 and 5.3.1 to find

$$\int_0^5 x \exp(-x^2) dx$$

(4) An integral that arises in the theory of blackbody radiation is

$$I(y) = \int_0^y f(x)\, dx, \text{ where } f(x) = \frac{x^3}{e^x - 1}, \text{ with } I(\infty) = \int_0^\infty \frac{x^3}{e^x - 1}\, dx = \frac{\pi^4}{15}.$$

Using the midpoint rule with $h = 0.1$ to compute $I(y)$, find the least value of y for which $I(y)$ is (a) 90%, (b) 99% of $I(\infty)$. Repeat with $h = 0.05$ to check the accuracy of your results. Note that, since

$$\lim_{x \to 0} \frac{x^3}{e^x - 1} = 0$$

it is safe put $f(0) = 0$ directly in the tabulation of $f(x)$, thus avoiding the error condition that would result if a spreadsheet tries to evaluate $f(0)$ by substituting $x = 0$ into $f(x) = x^3 / (e^x - 1)$.

§5.6* Romberg Integration

The trapezoidal rule produces results for which the error is $O(h^2)$. A simple manipulation of this rule can provide an "acceleration", resulting in a rule that is $O(h^4)$. In fact a scheme can be devised that is based on the trapezoidal rule and is capable of having error that is $O(h^n)$, where n is any (positive) multiple of 2.

The trapezoidal rule (5.4.2) can be written as

$$I = \int_a^b f(x)dx = \frac{h}{2}\left[f(a) + f(b) + 2\sum_{j=1}^{N-1} f(a+jh) \right] - \frac{Nh^3}{12} f''(\xi)$$

and it can be shown that if f has continuous derivatives (up to the 4th derivative, at least) then the trapezoidal rule has the form

$$I = \int_a^b f(x)dx = \frac{h}{2}\left[f(a) + f(b) + 2\sum_{j=1}^{N-1} f(a+jh) \right] - \frac{h^2}{12}[f'(b) - f'(a)]$$

$$+ \frac{(b-a)h^4}{720} f^{(4)}(\xi_h) \tag{5.6.1}$$

where $h = (b-a)/N$ and $\xi_h \in [a,b]$. For convenience we will write this as

$$I = I(h) + Ah^2 + Bf^{(4)}(\xi_h)h^4 \tag{5.6.2}$$

Here, $I(h)$ is the trapezoidal estimate of the integral for step-size h. If we now halve the step-size we get

$$I = I(\tfrac{1}{2}h) + \tfrac{1}{4}Ah^2 + \tfrac{1}{16}Bf^{(4)}(\xi_{\frac{1}{2}h})h^4 \tag{5.6.3}$$

If we now eliminate the $O(h^2)$ terms from equations 5.6.2 and 5.6.3 we find

$$I = \frac{4I(\tfrac{1}{2}h) - I(h)}{3} + \frac{Bh^4}{12}[f^{(4)}(\xi_{\frac{1}{2}h}) - f^{(4)}(\xi_h)] \tag{5.6.4}$$

In equation 5.6.4 we have an expression which enables us to get an estimate for I which is $O(h^4)$ from $I(h)$ and $I(\tfrac{1}{2}h)$ which are both $O(h^2)$. This "acceleration" is known as the Richardson extrapolation.

5.6.1 Example

Use the trapezoidal rule results found in Example 5.2.3 with $h = 0.2$ and $h = 0.1$ to find a more accurate estimate of the integral

$$I_1 = \int_0^1 x \exp(-x^2)dx = \tfrac{1}{2}(1 - \tfrac{1}{e}) \approx 0.3160603$$

From Example 5.2.3 we have:

$$I(0.2) = 0.31147897 \quad (|error| \approx 0.0046)$$

and

$$I(0.1) = 0.31491903 \quad (|error| \approx 0.0011)$$

and using the Richardson extrapolation we find

$$I \approx \frac{4I(0.1) - I(0.2)}{3} \approx 0.31606572 \text{ (with error } \approx 0.000005).$$

Note: The expression that results from the combination of equations 5.6.4 and 5.6.1 is none other than Simpson's rule!

Romberg integration is essentially the iterated application of the Richardson extrapolation. To explain this we need some notation:

Let $T(n,1)$ denote the trapezoidal estimate of the integral when $h_n = (b-a)/2^{n-1}$, and let $T(n,2)$ be the extrapolated values i.e.

$$T(n,2) = \frac{4T(n,1) - T(n-1,1)}{3}, \quad n = 2,3,4 \ldots \tag{5.6.5}$$

An extension of the trapezoidal rule of equation 5.6.1, assuming the derivatives of f up to the $(2n+2)^{nd}$ are continuous on $[a,b]$ leads the following generalization of (5.6.5):

$$T(i,j) = \frac{4^{j-1}T(i,j-1) - T(i-1,j-1)}{4^{j-1} - 1},$$

$$\text{where } i = 2,3,4,\ldots \text{ and } j = 2,3,\ldots,i-1,i \tag{5.6.6}$$

For example the second extrapolation $T(n,3)$ is given by

$$T(n,3) = \frac{16T(n,2) - T(n-1,2)}{15}$$

This process can be shown in a table like that below.

T(1,1)					
T(2,1)	T(2,2)				
T(3,1)	T(3,2)	T(3,3)			
T(4,1)	T(4,2)	T(4,3)	T(4,4)		
.	.	.	.	.	
.	.	.	.	.	
.	.	.	.	.	
T(n,1)	T(n,2)	T(n,3)	...		T(n,n)

The truncation error for $T(n,m)$ is $O(h_n^{2m})$.

5.6.2   Example

We revisit the integral of Example 5.2.3:

$$I_1 = \int_0^1 x \exp(-x^2)\,dx$$

and apply the Romberg method with $n = 5$.

It is not too difficult to implement Romberg integration in full on a spreadsheet, but here we simply substitute the $T(n,1)$ values (up to $n = 5$) into a structure like the table above.

The spreadsheet of Example 5.2.3 for the trapezoidal rule has been used to provide the $T(n,1)$ data in column B of the spreadsheet shown (in part) in Screen 5.6.1.

	A	B	C
1	Example 5.6.2	Romberg Integration	
2	\\ j =	=1	=B2+1
3	i	T(i,j)	T(i,j)
4	=1	=0.1839397206	
5	=A4+1	=0.2866700561	=(4^(C$2-1)*B5-B4)/(4^(C$2-1)-1)
6	=A5+1	=0.3088826241	=(4^(C$2-1)*B6-B5)/(4^(C$2-1)-1)
7	=A6+1	=0.3142758926	=(4^(C$2-1)*B7-B6)/(4^(C$2-1)-1)
8	=A7+1	=0.3156148009	=(4^(C$2-1)*B8-B7)/(4^(C$2-1)-1)

	D
2	=C2+1
3	T(i,j)
4	
5	
6	=(4^(D$2-1)*C6-C5)/(4^(D$2-1)-1)

	A	B	C
1	Example 5.6.2	Romberg Integration	
2	\\ j =	1	2
3	i	T(i,j)	T(i,j)
4	1	0.183939721	
5	2	0.286670056	0.320913501
6	3	0.308882624	0.316286813
7	4	0.314275893	0.316073649
8	5	0.315614801	0.316061104

	D	E	F
2	3	4	5
3	T(i,j)	T(i,j)	T(i,j)
4			
5			
6	0.315978368		
7	0.316059438	0.316060725	
8	0.316060267	0.316060280	0.316060279

Screen 5.6.1 *Romberg integration: extrapolation of trapezoidal rule to higher*

The formulas in cells C5..F8 correspond to the definition given in equation 5.6.6 for $T(i, j)$. The formulas not shown are easily created with the FILL command.

The truncation error in $T(5,1)$ is $O((\frac{1}{16})^2)$, whereas that in $T(5,5)$ is $O((\frac{1}{16})^{10})$, a significantly smaller value.

Exercises 5.6

(1) Apply Romberg integration to the integrals of Exercises 5.2 (1, 2 & 3), taking data generated by your spreadsheets for the trapezoidal rule in those exercises and substituting it into column B of the spreadsheet of Screen 5.6.1. Compare the error in $T(3,3)$, $T(4,4)$, and $T(5,5)$ with that resulting from the use of Simpson's rule with $N = 32$.

(2) Create a more general spreadsheet that can perform Romberg integration up to (say) $T(5,5)$ for any integrable function. This will require the inclusion of 5 separate trapezoidal routines, which could all be based on the same tabulation of the integrand function $f(x)$.

6 Ordinary Differential Equations

One of the most important applications of numerical methods is the solution of systems of one or more ordinary differential equations. In this chapter we first investigate some numerical methods for solving (for x as a function of t) the **initial-value problem** defined by the single first-order ordinary differential equation (ODE)

$$\frac{dx}{dt} = f(t,x), \quad 0 \le t \le T$$

with the **initial condition** $x = x(0)$ at $t = 0$.

Later we will consider systems of first-order equations, higher order equations and boundary-value problems.

We will assume that $x(0)$ and the function f are such that a unique solution exists. Details can be found in most relevant texts, many of which refer to the detailed study in Henrici[1962].

The differential equation is **autonomous** if f is not an explicit function of t. For example, the equation

$$\frac{dx}{dt} = \sin(x)$$

is autonomous, but

$$\frac{dx}{dt} = \exp(-t)\sin(x)$$

is not.

§6.1 The Euler and Modified Euler Methods

6.1.1 Subdivision of the Domain $[0, T]$

The interval $[0, T]$ is divided into N sub-intervals each of length $h = T / N$, and the continuous variable $x(t)$ is represented by the discrete set of values $x_0, x_1, x_2, ..., x_{N-1}, x_N$, where $x_n \approx x(t_n)$, $n = 0, 1, ..., N$ and $t_n = nh$, $n = 0, 1, 2, ..., N$. We begin with methods for solving the initial value problem

$$\frac{dx}{dt} = f(t, x), \quad 0 \le t \le T, \text{ with } x = x(0) \text{ at } t = 0 \qquad (6.1.1)$$

Notation: When a differential equation is solved using a numerical approximation then x_n and $x(t_n)$ will denote the approximated and exact values of x at $t = t_n$, respectively.

Methods for solving differential equations numerically can be classified as **one-step** or **Multi-step**. Only one-step methods are discussed in this book. These all have the form

$$x_{n+1} = x_n + h\Phi(t_n, x_n, h), \quad n = 0, 1, 2, ...$$

i.e. the value at the next point on the mesh depends only on the one preceding value - in contrast to multi-step methods where it depends on the values at two or more preceding steps.

Note: In the diagrams Fig. 6.1.1 and Fig. 6.1.2 that follow, for didactic purposes the point (t_n, x_n) is placed on the exact solution curve, implying that $x_n = x(t_n)$. In general this is only the case at the initial point x_0, unless the solution is a linear function of t. The numerical solution of (6.1.1) generates a polygonal approximation to the exact curve.

6.1.2 Euler's method

Taylor's theorem can be used to find x_{n+1} from x_n:

$$x_{n+1} = x(t_n + h) = x(t_n) + hx'(t_n) + h^2 x''(t_n) + \ldots$$
$$= x(t_n) + hf(t_n, x(t_n)) + \ldots \approx x_n + hf(t_n, x_n) + \ldots$$
$$\approx x_n + hf(t_n, x_n)$$

Euler's method is the difference equation and initial condition:

$$x_{n+1} = x_n + hf(t_n, x_n),\, n = 0, 1, 2, \ldots, N-1, \text{ with } x_0 = x(0) \quad (6.1.2)$$

Geometrically this amounts to replacing the graph of $x(t)$ with the tangent at t_n, as in Fig. 6.1.1, approximating $f(t_n, x(t_n))$ with $f(t_n, x_n)$, which is exact only at $n = 0$ if x is nonlinear. The point (x_n, t_n) actually lies on the polygonal approximating "curve", with "tangent" having gradient $f(t_n, x_n)$ there.

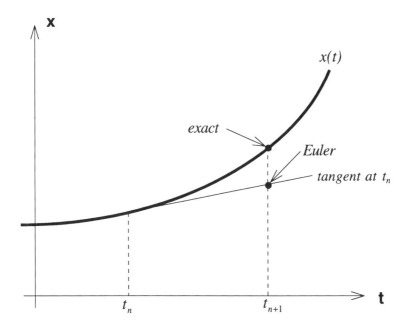

Fig. 6.1.1 *The geometrical basis for Euler's method: replace the curve with the tangent line at the current point.*

6.1.3 The Modified Euler method

A better approximation might be expected (see Fig. 6.1.2) if we take the average of the gradients found at each end of the sub-interval $[t_n, t_{n+1}]$, instead of just using the gradient at t_n , to predict the next point $x(t_{n+1})$. In this case

$$x_{n+1} \approx x_n + h \times \text{average slope}$$

$$= x_n + h \times \tfrac{1}{2}[f(t_n, x_n) + f(t_{n+1}, x_{n+1})]$$

(this is the *implicit* trapezoidal method referred to later, in §6.5.3).

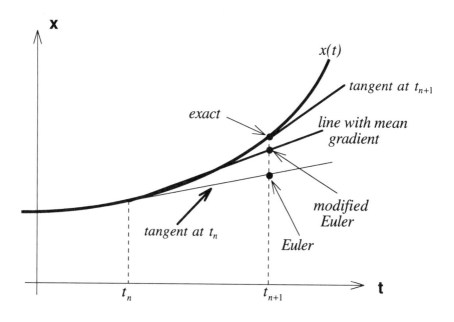

Fig. 6.1.2 *The modified Euler method uses the mean of the gradients at the beginning and end of the current interval, with the latter estimated using Euler's method.*

Now use the Euler result (6.1.2) to get the x_{n+1} term on the RHS, giving the *explicit* form (6.1.3), which is the modified Euler method.

Modified Euler method:

$$x_{n+1} \approx x_n + h \times \tfrac{1}{2}[f(t_n, x_n) + f(t_{n+1}, x_n + hf(t_n, x_n))] \qquad (6.1.3)$$
$$= x_n + \tfrac{1}{2}h[k_1(n) + k_2(n)]$$

where

$$k_1(n) = f(t_n, x_n) \text{ and } k_2(n) = f(t_{n+1}, x_n + hk_1(n))$$

6.1.4 Example

We begin with an initial-value problem that can be solved exactly:

$$\frac{dx}{dt} = 2\sqrt{x-1}, \quad x(0) = \alpha > 1, \text{ exact solution } x = (t + \sqrt{\alpha - 1})^2 + 1.$$

Screens 6.1.1 and 6.1.2 show the numerical solution of this problem using (a) the Euler and (b) modified Euler methods respectively, on the interval $0 \le t \le 10$, with $\alpha = 2$.

The exact solution, the error and the percentage error are also calculated, and the spreadsheet's graph plotting facility has been used to display the results, shown in Graphs 6.1.1, 6.1.2, 6.1.3, and 6.1.4.

Clearly, for this problem at least, the modified Euler method is the more accurate of the two.

The step-size h and initial value $x(0)$ are stored in cells D1 and D2 in both spreadsheets. The Euler method for this example is

$$x_{n+1} = x_n + h.2\sqrt{x_n - 1},$$

which is the sequence tabulated in cells C6, C7,... of Screen 6.1.1.

	A	B
1	Example	
2	6.1.4a	
3	Euler's method	
4	t(n)	x(n) exact
5	=0	=(A5+SQRT(D$2-1))^2+1
6	=A5+D$1	=(A6+SQRT(D$2-1))^2+1

	C	D	E
1	step-size h =	=0.2	
2	x(0) =	=2	
3			
4	x(n) Euler	% error	error
5	=D2	=100*(B5-C5)/B5	=B5-C5
6	=C5+D$1*2*SQRT(C5-1)	=100*(B6-C6)/B6	=B6-C6

	A	B	C	D	E
1	Example		step-size h =	0.2	
2	6.1.4a		x(0) =	2	
3	Euler's method				
4	t(n)	x(n) exact	x(n) Euler	% error	error
5	0	2	2.00	0.00	0.00
6	0.2	2.44	2.40	1.64	0.04
7	0.4	2.96	2.87	2.93	0.09

	A	B	C	D	E
54	9.8	117.64	112.61	4.28	5.03
55	10	122	116.83	4.24	5.17

Screen 6.1.1 *Solution of $\dot{x} = 2\sqrt{x-1}$, with $x(0) = \alpha > 1$, using Euler's method.*

For the modified Euler method in this example,

$$k_1(n) = 2\sqrt{x_n - 1}$$
$$k_2(n) = 2\sqrt{x_n + hk_1(n) - 1}$$

and these two functions are tabulated in columns D and E of Screen 6.1.2. The modified Euler solution is tabulated in cells C6, C7,... of that spreadsheet. The exact solution is calculated in column B in both cases.

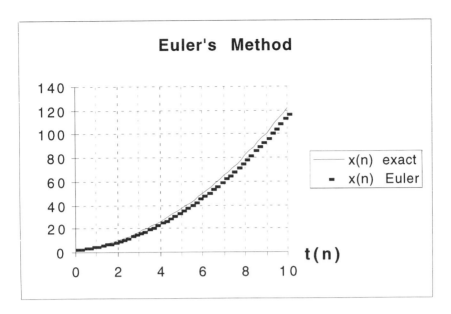

Graph 6.1.1 *Exact and Euler's method solutions, up to* t = *10, of the equation* $\dot{x} = 2\sqrt{x-1}$, *with* $x(0) = \alpha > 1$.

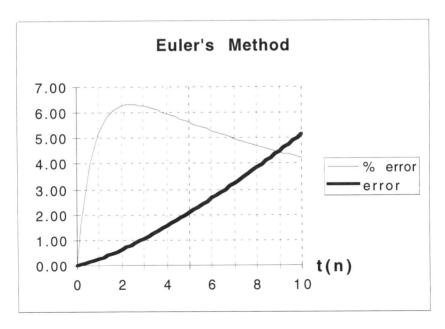

Graph 6.1.2 *The error in the Euler method.*

The decline of the percentage error toward zero occurs because the error rises roughly linearly while the solution rises quadratically. Exercises 6.1 will provide further insights into the accuracy of these methods.

	A	B
1	Example	
2	6.1.4b	
3	Modified	Euler
4	t(n)	x(n) exact
5	=0	=(A5+SQRT(D$2-1))^2+1
6	=A5+D$1	=(A6+SQRT(D$2-1))^2+1

	C	D	E
1	step-size h =	=0.2	
2	x(0) =	=2	
3	method		
4	x(n) M.E.	k1(n)	k2(n)
5	=D2	=2*SQRT(C5-1)	=2*SQRT(C5+D$1*D5-1)
6	=C5+0.5*D$1*(D5+E5)	=2*SQRT(C6-1)	=2*SQRT(C6+D$1*D6-1)

	F	G
4	% error	error
5	=100*(B5-C5)/B5	=B5-C5

	A	B	C	D	E	F	G
1	Example	step-size h =		0.2			
2	6.1.4b	x(0) =		2			
3	Modified	Euler	method				
4	t(n)	x(n) exact	x(n) M.E.	k1(n)	k2(n)	% error	error
5	0	2	2.00	2.000	2.366	0.00	0.00
6	0.2	2.44	2.44	2.397	2.768	0.14	0.00
7	0.4	2.96	2.95	2.795	3.170	0.23	0.01

	A	B	C	D	E	F	G
54	9.8	117.64	117.46	21.584	21.980	0.15	0.18
55	10	122	121.82	21.984	22.380	0.15	0.18

Screen 6.1.2 *Solution of* $\dot{x} = 2\sqrt{x-1}$, *with* $x(0) = \alpha > 1$ *using the modified Euler method*

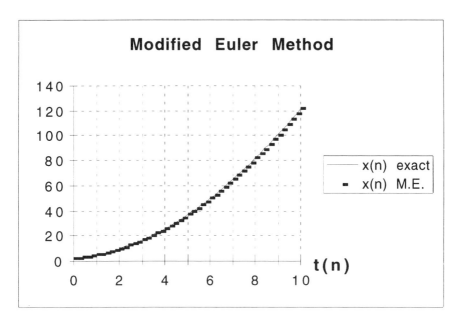

Graph 6.1.3 *Exact and modified Euler method solutions, up to t = 10, of the equation* $\dot{x} = 2\sqrt{x-1}$, *with* $x(0) = \alpha > 1$.

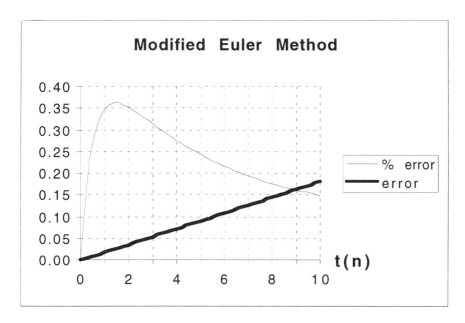

Graph 6.1.4 *Error in the modified Euler method solution.*

189

§6.2 Truncation (Discretization) Error and the Total Error

The error incurred when the differential equation is replaced with a finite difference equation is referred to as the **truncation** (or discretization) **error**. It is not the only source of error in the numerical solution of a differential equation, but it is usually the dominant one.

For the one-step method $x_{n+1} = x_n + h\Phi(t_n, x_n, h)$, $n = 0, 1, 2, \ldots$, the **local truncation error** d_{n+1} at t_{n+1} is defined by the difference between the exact value at t_{n+1} and the approximated value based on the exact value at the preceding mesh point:

$$d_{n+1} = x(t_{n+1}) - [x(t_n) + h\Phi(t_n, x(t_n), h)]$$

6.2.1 Definition: **Maximum Global Truncation Error $E(h)$**

If x_k and $x(t_k)$ are the values of the approximated and exact values (respectively) of $x(t)$ at $t = t_k$, then

$$E(h) = \underset{1 \leq k \leq N}{Max} |x_k - x(t_k)|$$

The **global truncation error** $g_n = x_n - x(t_n)$ is mainly determined by the local truncation error incurred at each step, and it can be shown that if the latter is $O(h^{n+1})$ then $E(h)$ will be $O(h^n)$. Such a method is called an nth-order method. Further details may be found in Schwarz[1989], for example.

For the Euler method, $E(h)$ is $O(h)$, and for the modified Euler method $E(h)$ is $O(h^2)$, so that for the latter case halving h will quarter the truncation error.

The modified Euler method is an example of a class of methods called (explicit) Runge-Kutta methods. Other explicit second-order Runge-Kutta methods are:

Heun's: $x_{n+1} = x_n + \frac{1}{4}h[f(t_n, x_n) + 3f(t_n + \frac{2}{3}h, x_n + \frac{2}{3}hf(t_n, x_n))]$

and Midpoint: $x_{n+1} = x_n + hf(t_n + \frac{1}{2}h, x_n + \frac{1}{2}hf(t_n, x_n))$

The **total error** in $x(T)$ will be the sum of the truncation error and the accumulated rounding error. The latter will arise from the evaluation of $f(t_n, x_n)$ and from the other calculations required to execute the chosen method. The truncation error approaches zero as h does, but smaller h implies a larger number of steps to reach a given T, and therefore a potentially greater accumulated rounding error. Thus there exists an optimum value of h below which it is pointless to reduce h any further. For many calculations the high precision of modern computers allows use of h values well above this optimum value.

Exercises 6.2

(1) Use the spreadsheet of Screen 6.1.2 to see how the error is reduced (for a given value of t) when h is halved, and halved again. You may need to increase the number of steps computed.

(2) Use the Euler, modified Euler and Heun methods to solve the following initial-value problems:

(a) $\quad \dfrac{dx}{dt} = \cos(t) + e^{-t}$, with $x(0) = 0$ for $0 \le t \le 1$,

(b) $\quad \dfrac{dx}{dt} = -x + t + 1$, with $x(0) = 1$ for $0 \le t \le 1$,

(c) $\quad \dfrac{dx}{dt} = -10x$, with $x(0) = 1$ for $0 \le t \le 5$,

(d) $\quad \dfrac{dx}{dt} = -5(x - \sin(t)) + \cos(t)$, with $x(0) = 1$, for $0 \le t \le 1.5$

The exact solutions are (a) $x(t) = \sin(t) - e^{-t} + 1$, (b) $x(t) = t + e^{-t}$, (c) $x(t) = e^{-10t}$, and (d) $x(t) = e^{-5t} + \sin(t)$.

Begin with $h = 0.2$ and then vary h, noting the effect on the error.

§6.3 The Fourth-Order Runge-Kutta Method

The modified Euler method gives a prediction of $x(t_{n+1})$ from $x(t_n)$ that uses a slope value that is the average of the exact slope at t_n and the estimated slope at t_{n+1}. A more subtle averaging process gives an extremely accurate method, one for which the global truncation error is $O(h^4)$. It is the

---

Fourth-order Runge-Kutta method:

$$x_{n+1} = x_n + \tfrac{1}{6}h[k_1(n) + 2k_2(n) + 2k_3(n) + k_4(n)] \qquad (6.3.1)$$

where

$$k_1(n) = f(t_n, x_n)$$
$$k_2(n) = f(t_n + \tfrac{1}{2}h, x_n + \tfrac{1}{2}hk_1(n))$$
$$k_3(n) = f(t_n + \tfrac{1}{2}h, x_n + \tfrac{1}{2}hk_2(n))$$
$$k_4(n) = f(t_n + h, x_n + hk_3(n))$$

6.3.1 Example

Apply the 4th order Runge-Kutta method to Example 6.1.4.

In this case we have

$$k_1(n) = f(t_n, x_n) = 2\sqrt{x_n} - 1$$
$$k_2(n) = 2\sqrt{x_n + \tfrac{1}{2}hk_1(n)} - 1$$
$$k_3(n) = 2\sqrt{x_n + \tfrac{1}{2}hk_2(n)} - 1$$
$$k_4(n) = 2\sqrt{x_n + hk_3(n)} - 1$$

	A	B
1	Example	
2	6.3.1	
3	4th	order
4		exact
5	t(n)	x(n)
6	=0	=(A6+SQRT(D$2-1))^2+1
7	=A6+D$1	=(A7+SQRT(D$2-1))^2+1

	C	D
1	step-size h =	=0.2
2	x(0) =	=2
3	Runge-	Kutta
4	R-K	
5	x(n)	k1(n)
6	=D2	=2*SQRT(C6-1)
7	=C6+D$1*(D6+2*E6+2*F6+G6)/6	=2*SQRT(C7-1)

	E	F
5	k2(n)	k3(n)
6	=2*SQRT(C6+0.5*D$1*D6-1)	=2*SQRT(C6+0.5*D$1*E6-1)

	G	H	I
5	k4(n)	% error	error
6	=2*SQRT(C6+D$1*F6-1)	=100*(B6-C6)/B6	=B6-C6
7	=2*SQRT(C7+D$1*F7-1)	=100*(B7-C7)/B7	=B7-C7

	A	B	C	D	E	F	G	H	I
1	Example	step-size h =	0.2						
2	6.3.1		x(0) =	2					
3		4th	order	Runge-	Kutta				
4			exact	R-K					
5	t(n)	x(n)	x(n)	k1(n)	k2(n)	k3(n)	k4(n)	% error	error
6	0	2	2.000	2.000	2.191	2.208	2.401	0.0E+00	0.0E+00
7	0.2	2.44	2.440	2.400	2.592	2.607	2.801	4.8E-04	1.2E-05

Screen 6.3.1 *Solution of a differential equation using the 4th order Runge-Kutta method.*

See Screen 6.3.1. These functions are tabulated in columns D, E, F and G, with the exact and Runge-Kutta solutions in columns B and C, respectively.

Graph 6.3.1 for the case $h = 0.2$ shows the percentage error to have magnitude less than about 0.00085%, an improvement by about a factor of more than 400 over the modified Euler method. The larger number of evaluations of f per step (greater potential rounding error) is more than adequately compensated for by the greatly reduced truncation error.

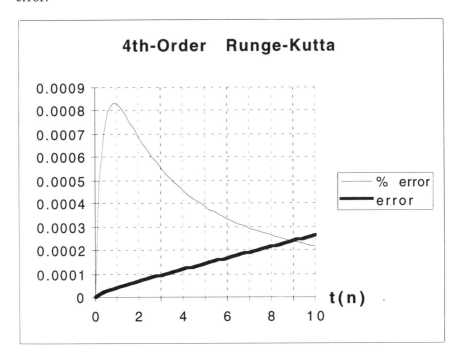

Graph 6.3.1

Exercises 6.3

Apply the 4th-order Runge-Kutta method to the problems listed in Exercise 6.2(2).

§6.4 Connections with Numerical Integration

If $\dfrac{dx}{dt} = f(t, x(t))$ then, by the fundamental theorem of calculus,

$$x_{n+1} - x_n = \int_{t_n}^{t_{n+1}} f(s, x(s))\, ds$$

Now use various approximations for the integral on the right-hand side:

(a) The left-hand end rule, $\displaystyle\int_{t_n}^{t_{n+1}} f(s, x(s))ds \approx hf(t_n, x(t_n)) \approx hf(t_n, x_n)$,

which gives Euler's method, or (b) the midpoint rule,

$$\int_{t_n}^{t_{n+1}} f(s, x(s))ds \approx hf(t_n + \tfrac{1}{2}h, x(t_n + \tfrac{1}{2}h)) \approx hf(t_n + \tfrac{1}{2}h, x_n + \tfrac{1}{2}hf(t_n, x_n)),$$

giving the midpoint method, or (c) the trapezoidal rule,

$$\int_{t_n}^{t_{n+1}} f(s, x(s))ds \approx \tfrac{1}{2}h(f(t_n, x_n) + f(t_{n+1}, x_{n+1})),$$

leading to the implicit trapezoidal method and the modified Euler method, or (d) Simpson's rule,

$$\int_{t_n}^{t_{n+1}} f(s, x(s))ds \approx \tfrac{1}{6}h[f(t_n, x_n) + 4f(t_{n+\frac{1}{2}}, x_{n+\frac{1}{2}}) + f(t_{n+1}, x_{n+1})]$$

which gives a result that *roughly* resembles the 4th-order Runge-Kutta method. The marked superiority of the latter over the modified Euler method reflects the significant improvement in accuracy afforded by Simpson's rule compared with the trapezoidal rule. A closer examination of the connection between numerical quadrature and numerical solution of differential equations can be found in the literature, for example in Schwarz [1989].

§6.5 Stability

There are various aspects of the concept of stability to be considered in relation to the solution of differential equations. A simplified discussion follows below.

First there is the notion of inherent instability in the sense of the *problem* being ill-conditioned (or not well-posed). An initial-value problem is ill-conditioned if a small change in the initial condition or in a parameter leads to a large change in the solution. Further, it is desirable that the numerical method should be stable in the sense that a small variation in the initial data should lead to a small variation in the solution, i.e. that the (numerical) solution should depend *continuously* on the initial data. Likely sources of such variations are the errors inherent in experimentally determined quantities, and the inability of computers to represent some numbers exactly.

Many numerical analysts prefer not to classify ill-conditioning (a deficiency of the problem) as a type of instability, reserving this term for properties of the method of solution.

If a problem is well-posed, there remains the possibility of instability in the *method* used to solve it numerically, for instance the presence of a spurious exponentially growing term, referred to as *induced* instability. The most likely source of such a term is the truncation error resulting from the replacement of the differential equation with a difference scheme.

6.5.1 Example

The problem

$$\frac{dx}{dt} = x - t, \quad x(0) = 1$$

has exact solution $x = t + 1$, but if the initial condition is changed to $x(0) = 1 + \varepsilon$ (where ε is small), the exact solution becomes $x = \varepsilon e^t + t + 1$ which behaves very differently for large t. This problem is certainly ill-conditioned, or inherently unstable.

6.5.2 Example

We consider the use of the modified Euler method to solve the problem of Exercise 6.2.2(c):

$$\frac{dx}{dt} = -10x \text{ with } x(0) = 1, \text{ exact solution } x(t) = e^{-10t} .$$

The modified Euler method gives the difference equation

$$x_{n+1} = x_n + \tfrac{1}{2}h[-10x_n - 10(x_n - 10hx_n)] = x_n[1 - 10h + 50h^2]$$

which has solution

$$x_n = x_0[1 - 10h + 50h^2]^n = [1 - 10h + 50h^2]^n.$$

The worth of this solution is measured by how well it approximates the exact solution, in this case an exponentially decaying transient. If h is small we have a good approximation to the exact solution, but on the other hand if $|1 - 10h + 50h^2| > 1$ we have a solution which *grows* exponentially. In this case we need $h < 0.2$ to ensure that the numerical solution should (at least) not exhibit this exponential growth.

Note: As $h \to 0$ (and $N = T/h \to \infty$), $x_n \to \exp(-10t_n)$, i.e. for small enough h we get good answers.

6.5.3 Stability Criteria

In Example 6.5.2 we applied the modified Euler method to the equation $x' = -10x$ and found that $h < 0.2$ was required for a stable solution. For the equation $x' = \lambda x$ the modified Euler method gives

$$x_{n+1} = \left(1 + \lambda h + \tfrac{1}{2}(\lambda h)^2\right)x_n$$

which has solution $x_n = \left(1 + \lambda h + \tfrac{1}{2}(\lambda h)^2\right)^n x_0$. Stability dictates that $\left|1 + \lambda h + \tfrac{1}{2}(\lambda h)^2\right| < 1$ when this method is used. As we are only interested in $\lambda < 0$ this condition is $h|\lambda| < 2$.

The 4th-order Runge-Kutta method gives the difference equation

$$x_{n+1} = \left(1 + \lambda h + \tfrac{1}{2}(\lambda h)^2 + \tfrac{1}{6}(\lambda h)^3 + \tfrac{1}{24}(\lambda h)^4\right)x_n,$$

for which stability requires that

$$\left|1 + \lambda h + \tfrac{1}{2}(\lambda h)^2 + \tfrac{1}{6}(\lambda h)^3 + \tfrac{1}{24}(\lambda h)^4\right| < 1,$$

which is satisfied for $-2.78 < h\lambda < 0$ i.e. $h|\lambda| < 2.78$, approximately.

In fact the explicit p^{th}-order Runge-Kutta methods all have the form

$$x_{n+1} = F(\lambda h)x_n,$$

where $F(\lambda h)$ is the first $p+1$ terms of the Taylor series for $\exp(\lambda h)$. The accuracy obtained when solving $x' = \lambda x$ thus depends on how well $F(\lambda h)$ represents $\exp(\lambda h)$.

In §6.1 the first stage of the derivation of the modified Euler method (before substitution for x_{n+1} using the Euler approximation) actually results in another method, the **implicit trapezoidal method**, which is the iteration

$$x_{n+1} = x_n + \tfrac{1}{2}h[f(t_{n+1}, x_{n+1}) + f(t_n, x_n)]$$

Generally this equation cannot be solved explicitly for x_{n+1} in terms of x_n and thus a numerical method such as Newton's will be needed. However for some cases - such as when f is linear - an explicit rearrangement is possible.

If this method is applied to the equation $x' = \lambda x$ it gives

$$x_{n+1} = \frac{1 + \tfrac{1}{2}\lambda h}{1 - \tfrac{1}{2}\lambda h} x_n$$

for which the stability condition

$$\left| \frac{1+\frac{1}{2}\lambda h}{1-\frac{1}{2}\lambda h} \right| < 1$$

is satisfied for *all* h (assuming $\lambda < 0$).

The point of this discussion is to indicate the fact that some methods allow a less restricted choice of step-size, at the possible price of more complex computations. Each method has its "zone of stability", the range of values of λh for which it is stable. This matter will be discussed further when we consider stiff systems in §6.10.

Of course most differential equations of interest are nonlinear, yet the stability classification of the various numerical methods using the "standard" linear equation $x' = \lambda x$ still applies. This is because the local behaviour over one small step will be governed by the linear part of the Taylor expansion of the nonlinear equation.

Exercises 6.5

(1) For each of the following, set up a spreadsheet that tabulates the exact solution, the modified Euler and the implicit trapezoidal approximate solutions:

(a) $x' = -10x$ with $x(0) = 1$, exact solution $x(t) = e^{-10t}$,

(b) $x' = -10(x-t)+1$ with $x(0) = 1$, exact solution $x(t) = e^{-10t} + t$,

(c) $x' = -10(x - \sin(\pi t)) + \pi\cos(\pi t)$ with $x(0) = 1$, which has exact solution $x(t) = e^{-10t} + \sin(\pi t)$,

and compare them at $t = 1$ for $h = 0.01, 0.02, 0.05, 0.1, 0.25$, etc. Tabulation of the percentage errors in the numerical solutions will help the comparison, as will suitable graphs.

(2) Solve the problems above using the 4th-order Runge-Kutta method, again comparing with the other methods as h is increased.

§6.6 Coupled First-Order Differential Equations

The methods studied above for solving a single 1st-order ODE are all readily extended to a system of two or more coupled 1st-order ODE's, for example the (second-order) system represented by the pair of first-order equations

$$\frac{dx}{dt} = f(t, x, y)$$

$$\frac{dy}{dt} = g(t, x, y)$$

(6.6.1)

with given initial conditions $x(0)$ and $y(0)$ at $t = 0$.

To solve (6.6.1) numerically for $0 \leq t \leq T$, first subdivide the interval $[0, T]$ into N sub-intervals of length $h = T / N$. Take $x_0 = x(0)$, $y_0 = y(0)$ and, for $n = 0, 1, 2, ..., N$, use the difference equations (6.6.2) or (6.6.3) given below for the two methods considered here.

6.6.1 Modified Euler Method

$$x_{n+1} = x_n + \tfrac{1}{2} h[k_1(n) + k_2(n)]$$

$$y_{n+1} = y_n + \tfrac{1}{2} h[l_1(n) + l_2(n)]$$

(6.6.2)

where

$$k_1(n) = f(t_n, x_n, y_n)$$

$$k_2(n) = f(t_{n+1}, x_n + hk_1(n), y_n + hl_1(n))$$

$$l_1(n) = g(t_n, x_n, y_n)$$

$$l_2(n) = g(t_{n+1}, x_n + hk_1(n), y_n + hl_1(n))$$

6.6.3 Example

Apply the modified Euler method to the pair of equations

$$\frac{dx}{dt} = f(t,x,y) = x+y, \qquad x(0) = 0$$

$$\frac{dy}{dt} = g(t,x,y) = -x+y, \qquad y(0) = 1$$

The exact solution is $x(t) = e^t \sin(t)$, $y(t) = e^t \cos(t)$.

It follows that

$$k_1(n) = x_n + y_n$$
$$k_2(n) = x_n + hk_1(n) + y_n + hl_1(n))$$
$$l_1(n) = -x_n + y_n$$
$$l_2(n) = -(x_n + hk_1(n)) + y_n + hl_1(n)$$

Screen 6.6.1 shows the spreadsheet for this problem. The approximated x_n and y_n are listed in columns B and C, the functions $k_1(n), k_2(n), l_1(n)$, and $l_2(n)$ are in columns D, E, F and G respectively, and the exact x and y values in columns H and I. The percentage proportional error and the errors in y and y are in columns J, K and L.

Graph 6.6.1 shows a plot of x vs y for exact and numerical solutions, and Graph 6.6.2 shows the errors in x and y and the percentage proportional error $P(t_n)$, defined by

$$P(t_n) = 100 \times \frac{\text{distance from exact point to approximate point}}{\text{distance from exact point to origin}}$$

$$= 100 \sqrt{\frac{(x(t_n) - x_n)^2 + (y(t_n) - y_n)^2}{x(t_n)^2 + y(t_n)^2}}$$

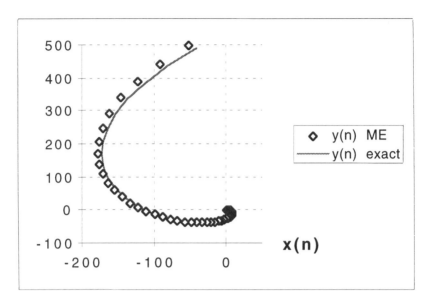

Graph 6.6.1 Exact and numerical solutions plotted in the xy plane.

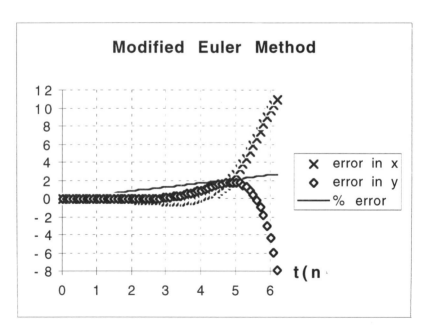

Graph 6.6.2 Plots of errors in x and y, and of percentage proportional error.

	A	B	C	D
1	Example		step-size h =	=0.1
2	6.6.3		x(0) =	=0
3	Mod.	Euler Method	y(0) =	=1
4	2	coupled	ODE's	
5	t(n)	x(n) ME	y(n) ME	k1(n)
6	=0	=D2	=D3	=B6+C6
7	=A6+D$1	=B6+D$1*(D6+E6)/2	=C6+D$1*(F6+G6)/2	=B7+C7

	E	F	G
5	k2(n)	11(n)	12(n)
6	=B6+D$1*D6+C6+D$1*F6	=-B6+C6	=-(B6+D$1*D6)+C6+D$1*F6

	H	I
5	x(n) exact	y(n) exact
6	=EXP(A6)*SIN(A6)	=EXP(A6)*COS(A6)

	J	K	L
5	% error	error in x	error in y
6	=100*SQRT(K6^2+L6^2)/SQRT(H6^2+I6^2)	=H6-B6	=I6-C6

	A	B	C	D	E	F	G
1	Example		step-size h =	0.1			
2	6.6.3		x(0) =	0			
3	Mod.	Euler Method	y(0) =	1			
4	2	coupled	ODE's				
5	t(n)	x(n) ME	y(n) ME	k1(n)	k2(n)	11(n)	12(n)
6	0	0	1	1.0E+00	1.2E+00	1.0E+00	1.0E+00
7	0.1	0.11	1.1	1.2E+00	1.4E+00	9.9E-01	9.7E-01

	H	I	J	K	L
5	x(n) exact	y(n) exact	% error	error in x	error in y
6	0.000	1.000	0	0	0
7	0.110	1.100	0.04373	0.000333	-0.00035

Screen 6.6.1 *Solution of a pair of differential equations using the modified Euler method.*

The quantity $\sqrt{(x(t_n)-x_n)^2+(y(t_n)-y_n)^2}$ is the magnitude of the global error at $t=t_n$. In this problem involving exponential growth the proportional error gives a more realistic perspective to the error growth.

The data shown are for the case $h = 0.1$, with $0 \le t \le 2\pi \approx 6.3$. The percentage proportional error increases linearly to approximately 2.8% at $t = 6.3$. When h is reduced to 0.05, this goes down to about 0.71%, a reduction by a factor of about one quarter - as expected since the method is $O(h^2)$.

Another method for a pair of 1st-order ODE's is the 4th-order Runge-Kutta method.

4.4.2 Fourth-order Runge-Kutta Method

$$x_{n+1} = x_n + \tfrac{1}{6}h[k_1(n) + 2k_2(n) + 2k_3(n) + k_4(n)]$$
$$y_{n+1} = y_n + \tfrac{1}{6}h[l_1(n) + 2l_2(n) + 2l_3(n) + l_4(n)]$$

(6.6.3)

where

$$k_1(n) = f(t_n, x_n, y_n)$$
$$k_2(n) = f(t_{n+\frac{1}{2}}, x_n + \tfrac{1}{2}hk_1(n), y_n + \tfrac{1}{2}hl_1(n))$$
$$k_3(n) = f(t_{n+\frac{1}{2}}, x_n + \tfrac{1}{2}hk_2(n), y_n + \tfrac{1}{2}hl_2(n))$$
$$k_4(n) = f(t_{n+1}, x_n + hk_3(n), y_n + hl_3(n))$$

and

$$l_1(n) = g(t_n, x_n, y_n)$$
$$l_2(n) = g(t_{n+\frac{1}{2}}, x_n + \tfrac{1}{2}hk_1(n), y_n + \tfrac{1}{2}hl_1(n))$$
$$l_3(n) = g(t_{n+\frac{1}{2}}, x_n + \tfrac{1}{2}hk_2(n), y_n + \tfrac{1}{2}hl_2(n))$$
$$l_4(n) = g(t_{n+1}, x_n + hk_3(n), y_n + hl_3(n))$$

Exercises 6.6

(1) Use the modified Euler method to solve, for $0 \le x \le 6.2$,

$$\frac{dx}{dt} = f(t,x,y) = -x + y, \quad x(0) = 0$$
$$\frac{dy}{dt} = g(t,x,y) = -x - y, \quad y(0) = 1$$

In this case the exact solution is $x(t) = e^{-t} \sin(t)$, $y(t) = e^{-t} \cos(t)$, which converges to the origin for large t

(2) Apply the 4th-order Runge-Kutta method to (a) the problem of Example 6.6.3 , (b) to problem(1) above, and (c) to the system

$$\frac{dx}{dt} = y + x(1 - x^2 - y^2), \quad x(0) = 0.5$$
$$\frac{dy}{dt} = -x + y(1 - x^2 - y^2), \quad y(0) = 0$$

which has solution $x = \sqrt{1 + 3e^{-2t}} \cos(-t)$, $y = \sqrt{1 + 3e^{-2t}} \sin(-t)$.
Compare the accuracy with that obtained using modified Euler, particularly when larger maximum values of t are used.

(3) A simple biological model of a predator-prey system describing the time variation of the (scaled) numbers of predators $y(t)$ and prey $x(t)$ is the Lotka-Volterra equations:

$$\dot{x}(t) = ax(1 - y)$$
$$\dot{y}(t) = y(x - 1)$$

Solve this pair of equations using the 4th-order Runge-Kutta method, initially for the case $a = 10$, $x(0) = 4$, $y(0) = 1.5$, with step-size $h = 0.025$, for $0 \le t \le 5$.

(4) A simple model for the levels of sugar x and insulin y in the blood of non diabetic and diabetic humans is given by the following pair of coupled ODE's:

$$\frac{dx}{dt} = -a_1 xy + a_2(x_f - x)H(x_f - x) + z(t)$$

$$\frac{dy}{dt} = b_1(x - x_f)H(x - x_f) - b_2 y + w(t)$$

where

$$H(x - x_f) = \begin{cases} 1, & x > x_f \\ 0 & x \le x_f \end{cases}$$

In the first equation, the first term represents the reduction of sugar level in the presence of insulin, and the second term gives release of sugar from the liver when the level falls below the fasting level x_f. The third term $z(t)$ gives the increase in sugar level due to food intake.

The first term of the second equation gives an increase in insulin level due to secretion from the pancreas when the sugar rises above the fasting level. The second term repesents the natural decomposition of insulin (exponential decay with a half-life of about 10-25 minutes), and the final term $w(t)$ represents the increase in insulin due to injection. This last term is zero in the case of a non-diabetic.

Suitable parameter values that typify non-diabetic and diabetic metabolisms are as follows:

Type	a_1	a_2	b_1	b_2
non-diabetic	0.05	1	0.5	2
diabetic	0.01	1	0.01	2

The food (sugar) intake term $z(t)$ can be modelled as the sum of three terms $z_B(t)$, $z_L(t)$ and $z_T(t)$, representing the sugar intake at Breakfast, Lunch and Tea, respectively. At each meal, the sugar intake will be maximum just after the meal is consumed, falling off rapidly thereafter. This process can be modelled with the half Gaussian function:

$$z_M = \begin{cases} 0, & t < t_M \\ a_M \exp(-K(t - t_M)^2) & t \ge t_M \end{cases}$$

where M is B, L or T, a_M is a positive constant repesenting the size of the meal, t_M is the time of the meal, and K is a measure of the time it takes the sugar in the food to appear in the blood stream. We use $K = 2$ hr^{-1}.

The total sugar in meal M is given by

$$S(M) = a_M \int_0^\infty z_M(t)dt = \tfrac{1}{2}\sqrt{\frac{\pi}{K}}a_M$$

With a fasting sugar level $x_f = 100$ units, suitable times and sugar contents for the meals are:

Meal M	Time t_M	$S(M)$ (units)
Breakfast	8 am	50
Lunch	12 noon	75
Tea	6 pm	100

The periodic injections of insulin needed by a diabetic can be modelled in a similar way. The full effect of an injection is experienced after about three hours, and diminishes entirely after about 6 hours. This is most easily represented with a triangular pulse, symmetric about $t = T + 3$, where T is the time of injection. The area under the graph of a pulse equals the total insulin given in that injection.

(a) Use the modified Euler method to solve the coupled pair of ODE's for the case of a non-diabetic, for $6 \le t \le 24$, with t step-size $h = 0.15$ hours. Plot the sugar and insulin levels versus time.

(b) Change the parameters a_1 and b_1 to obtain a simulation of a diabetic who is not receiving insulin injections. Note the uncontrolled increase in sugar level.

(c) Include two injections of insulin, one of 100 units at 07.30 hrs., and one 220 units at 15.30 hrs. The first is intended to aid sugar metabolism for both breakfast and lunch, and the second for the evening meal. The result should be better control of sugar level over the day.

The times and sizes of the meals and injections should be convenient for variation, for example by putting them at the top of the spreadsheet, above the main calculations.

§6.7 Higher Order ODE's and First-Order Equations

One approach to the solution of a higher order Ordinary Differential Equation is to first convert it to an equivalent system of 1st-order ODE's (if possible) and then proceed as in the previous section. Thus if we have the mth-order equation

$$x^{(m)}(t) = \frac{d^m x}{dt^m} = f(t, x, x', x'', \ldots, x^{(m-1)}) \tag{6.7.1}$$

with the initial conditions

$$x(0) = a_1, \, x'(0) = a_2, \, x''(0) = a_3, \ldots, x^{(m-1)}(0) = a_m,$$

we can define the new variables $y_k(t) = x^{(k-1)}(t)$, $k = 1, 2, \ldots, m$, for which the initial conditions become $y_k(0) = a_k$, $k = 1, 2, \ldots, m$.

For example, $y_1(t) = x(t)$, $y_2(t) = x'(t)$, and $y_1(0) = x(0) = a_1$, $y_2(0) = x'(0) = a_2$.

With these changes the mth-order equation (6.7.1) above is equivalent to the system

$$\frac{dy_1}{dt} = y_2$$

$$\frac{dy_2}{dt} = y_3$$

$$\vdots \tag{6.7.2}$$

$$\frac{dy_{m-1}}{dt} = y_m$$

$$\frac{dy_m}{dt} = f(t, y_1, y_2, \ldots, y_m)$$

6.7.1 Example

Express the following second-order initial-value problem as a pair of first order equations with appropriate initial conditions:

$$\frac{d^2x}{dt^2} + \left(\frac{dx}{dt}\right)^2 - 4x = \sin(t), \text{ with } x(0)=1, x'(0)=2.$$

First, define $y_1 = x$ and

$$y_2 = \frac{dx}{dt} = \frac{dy_1}{dt} \Rightarrow \frac{dy_2}{dt} = \frac{d^2x}{dt^2}.$$

The second-order initial-value problem above now becomes

$$\frac{dy_1}{dt} = y_2, \qquad\qquad y_1(0)=1$$
$$\frac{dy_2}{dt} = -y_2^2 + 4y_1 + \sin(t), \quad y_2(0)=2$$

6.7.2 Example

The second-order initial-value problem

$$\frac{d^2x}{dt^2} + 2\frac{dx}{dt} + 2x = R\sin(t) \text{ with } x(0)=0=x'(0)$$

describes a damped harmonic oscillator driven at its resonant frequency by a sinusoidal source of amplitude R, starting from rest at its equilibrium position. Express this equation as a pair of first-order *equations and solve them numerically using the 4th-order Runge-Kutta method. The exact solution (with $\phi = \tan^{-1}(2)$) is*

$$x(t) = \frac{R}{\sqrt{5}}[e^{-t}\sin(t+\phi) + \sin(t-\phi)]$$

$$y(t) = \frac{R}{\sqrt{5}}[e^{-t}(\cos(t+\phi) - \sin(t+\phi)) + \cos(t-\phi)]$$

The problem can be rewritten as the first-order pair

$$\frac{dx}{dt} = y, \qquad\qquad x(0) = 0$$

$$\frac{dy}{dt} = -2x - 2y + R\sin(t), \quad y(0) = 0$$

The variable x is the displacement from equiluilibrium and y is the velocity at time t. In Screen 6.7.1 we apply the 4th-order Runge-Kutta method to this problem, solving it on the interval $0 \le t \le 15$ with $h = 0.1$ and $R = 1$. The global error does not exceed 2.5×10^{-6}, up to $t = 15$. Graphs 6.7.1 and 6.7.2 show x and y vs t and the global error vs t, respectively.

Applying equations (6.6.3) to this problem, we have

$$k_1(n) = y_n$$
$$k_2(n) = y_n + \tfrac{1}{2}hl_1(n)$$
$$k_3(n) = y_n + \tfrac{1}{2}hl_2(n)$$
$$k_4(n) = y_n + hl_3(n)$$

and

$$l_1(n) = -2x_n - 2y_n + R\sin(t_n)$$
$$l_2(n) = -2(x_n + \tfrac{1}{2}hk_1(n)) - 2(y_n + \tfrac{1}{2}hl_1(n)) + R\sin(t_{n+\frac{1}{2}})$$
$$l_3(n) = -2(x_n + \tfrac{1}{2}hk_2(n)) - 2(y_n + \tfrac{1}{2}hl_2(n)) + R\sin(t_{n+\frac{1}{2}})$$
$$l_4(n) = -2(x_n + hk_3(n)) - 2(y_n + hl_3(n)) + R\sin(t_{n+1})$$

	A	B
1	Example 6.7.2	
2	Runge-Kutta method	
3	two coupled ODE's	
4	Driven damped oscillator	
5	t(n)	x(n) RK
6	=0	=D2
7	=A6+D$1	=B6+D$1*(D6+2*E6+2*F6+G6)/6

	C	D	E
1	step-size h =	=0.1	phi =
2	x(0) =	=0	
3	y(0) =	=0	
4	R =	=1	
5	y(n) RK	k1(n)	k2(n)
6	=D3	=C6	=C6+D$1*H6/2
7	=C6+D$1*(H6+2*I6+2*J6+K6)/6	=C7	=C7+D$1*H7/2

	F	G	H
1	=ATAN(2)		
2			
3			
4			
5	k3(n)	k4(n)	l1(n)
6	=C6+D$1*I6/2	=C6+D$1*J6	=-2*B6-2*C6+D$4*SIN(A6)

	I
5	l2(n)
6	=-2*(B6+D$1*D6/2)-2*(C6+D$1*H6/2)+D$4*SIN(A6+D$1/2)

	J
5	l3(n)
6	=-2*(B6+D$1*E6/2)-2*(C6+D$1*I6/2)+D$4*SIN(A6+D$1/2)

	K
5	l4(n)
6	=-2*(B6+D$1*F6)-2*(C6+D$1*J6)+D$4*SIN(A6+D$1)

Screen 6.7.1 *Solution of a pair of 1st order ODE's (equivalent to a 2nd order ODE) using the 4th-order Runge-Kutta method. The equations model a damped oscillator being driven at its resonant frequency.*

	L
5	**x(n) exact**
6	=D$4*(EXP(-A6)*SIN(A6+F$1)+SIN(A6-F$1))/SQRT(5)

	M
5	**y(n) exact**
6	=D$4*(EXP(-A6)*(COS(A6+F$1)-SIN(A6+F$1))+COS(A6-F$1))/SQRT(5)

	N	O	P	Q
5	**|global error|**	**error in x**	**error in y**	**% prop error**
6	=SQRT(O6^2+P6^2)	=L6-B6	=M6-C6	
7	=SQRT(O7^2+P7^2)	=L7-B7	=M7-C7	=100*N7/SQRT(B7^2+C7^2)

	A	B	C	D	E	F	G
1	Example 6.7.2		step-size h =	0.1		phi =	1.10715
2	Runge-Kutta method		x(0) =	0			
3	two coupled ODE's		y(0) =	0			
4	Driven damped oscillator		R =	1			
5	t(n)	x(n) RK	y(n) RK	k1(n)	k2(n)	k3(n)	k4(n)
6	0	0	0	0.0E+00	0.0E+00	2.5E-03	4.5E-03
7	0.1	0.000158267	0.00467097	4.7E-03	9.2E-03	1.1E-02	1.7E-02

	H	I	J	K	L	M
5	l1(n)	l2(n)	l3(n)	l4(n)	x(n) exact	y(n) exact
6	0.0E+00	5.0E-02	4.5E-02	9.0E-02	0.000E+00	9.930E-17
7	9.0E-02	1.3E-01	1.3E-01	1.6E-01	1.584E-04	4.671E-03

	N	O	P	Q
5	**|global error|**	**error in x**	**error in y**	**% prop error**
6	9.9E-17	0.0E+00	9.9E-17	
7	1.5E-07	1.5E-07	2.3E-08	0.003288431

Screen 6.7.1 (cont)

The functions $k_1(n),...,l_4(n)$ are tabulated in columns D, E,...,K respectively, the R-K approximated solutions for x and y are in columns B and C, and the exact values are in columns L and M. The global error and the errors in x and y are in columns N, O and P.

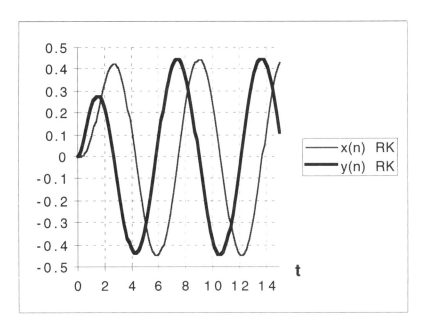

Graph 6.7.1 *A plot of the displacement x and velocity y against time, using Runge-Kutta.*

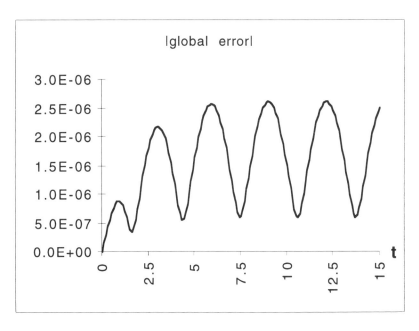

Graph 6.7.2 *Plot of the magnitude of the global error against t.*

Exercises 6.7

(1) Use the modified Euler method to solve the problem of Example 6.7.2.

(2) An important equation in electronics is Van der Pol's equation describing a relaxation oscillator, $\ddot{x} - \varepsilon(1 - x^2)\dot{x} + ax = 0$, equivalent to the 1st-order pair

$$\dot{x} = y$$

$$\dot{y} = -ax + \varepsilon(1 - x^2)y$$

Solve this system using the 4th-order Runge-Kutta method with step-size 0.1, for the case $a = 2$, $\varepsilon = 2.5$, $x(0) = 1.5$, $y(0) = 1$.

§6.8 Finite Difference Approximations for ODE's

Let us apply Taylor's theorem to the continuous function $x(t)$ at an interior point of the mesh, $n = 1, 2, ..., N - 1$. Then

$$x_{n+1} = x(t_n + h)$$
$$= x(t_n) + hx'(t_n) + \tfrac{1}{2}h^2 x''(t_n) + \tfrac{1}{3!}h^3 x^{(3)}(t_n) + \tfrac{1}{4!}h^4 x^{(4)}(\xi_n^+)$$
$$= x_n + hx'_n + \tfrac{1}{2}h^2 x''_n + \tfrac{1}{3!}h^3 x_n^{(3)} + \tfrac{1}{4!}x^{(4)}h^4(\xi_n^+)$$

and

$$x_{n-1} = x(t_n - h)$$
$$= x(t_n) - hx'(t_n) + \tfrac{1}{2}h^2 x''(t_n) - \tfrac{1}{3!}h^3 x^{(3)}(t_n) + \tfrac{1}{4!}h^4 x^{(4)}(\xi_n^-)$$
$$= x_n - hx'_n + \tfrac{1}{2}h^2 x''_n - \tfrac{1}{3!}h^3 x_n^{(3)} + \tfrac{1}{4!}h^4 x^{(4)}(\xi_n^-)$$

where $x_n < \xi_n^+ < x_{n+1}$ and $x_{n-1} < \xi_n^- < x_n$. If the two expressions are added we get

$$x_{n+1} + x_{n-1} = 2x_n + h^2 x''_n + \tfrac{1}{4!}h^4[x^{(4)}(\xi_n^-) + x^{(4)}(\xi_n^+)]$$

and the Intermediate Value Theorem allows the last term to be simplified so that we have

$$x_n'' = \frac{x_{n+1} - 2x_n + x_{n-1}}{h^2} - \frac{1}{12}h^2 x^{(4)}(\xi_n), \text{ with } x_{n-1} < \xi_n < x_{n+1}.$$

A similar exercise where the two Taylor expansions are subtracted leads to

$$x_n' = \frac{x_{n+1} - x_{n-1}}{2h} - \frac{1}{6}h^2 x^{(3)}(\eta_n), \text{ with } x_{n-1} < \eta_n < x_{n+1}.$$

We now have the **central difference approximations** for the first and second derivatives of x, both of them being $O(h^2)$, namely

$$x_n' \approx \frac{x_{n+1} - x_{n-1}}{2h} \text{ and } x_n'' \approx \frac{x_{n+1} - 2x_n + x_{n-1}}{h^2} \qquad (6.8.1)$$

There are other approximations having lower or higher accuracy but these two are of particular interest to us here.

Application to Linear Initial-Value Problems

6.8.1 Example

Find a finite difference approximation for the second-order initial-value problem

$$\frac{d^2x}{dt^2} + x^2\frac{dx}{dt} - x = e^{-t}, \text{ with } x(0) = 2, x'(0) = 1.$$

Using the approximations (6.8.1) for the derivatives found above the ODE has the finite difference approximation

$$\frac{x_{n+1} - 2x_n + x_{n-1}}{h^2} + x_n^2 \frac{x_{n+1} - x_{n-1}}{2h} - x_n = \exp(-t_n)$$

and the initial conditions are transformed as follows:

$$x(0) = 2 \Rightarrow x_0 = 2, \quad x'(0) = 1 \Rightarrow x_0' \approx \frac{x_1 - x_{-1}}{2h} = 1 \Rightarrow x_1 - x_{-1} = 2h.$$

Putting $n = 0$ in the difference equation gives

$$\frac{x_1 - 2x_0 + x_{-1}}{h^2} + x_0^2 \frac{x_1 - x_{-1}}{2h} - x_0 = \exp(-t_0) = 1$$

$$\Rightarrow x_1 - 2x_0 + x_1 - 2h + h^2 x_0^2 - h^2 x_0 = h^2$$

Along with $x_0 = 2$ this gives $x_1 = 2 + h - \frac{1}{2}h^2$. The original problem is now approximated by the difference equation

$$x_{n+1} = \frac{(2 + h^2)x_n + (\frac{1}{2}hx_n^2 - 1)x_{n-1} + h^2 \exp(-t_n)}{1 + \frac{1}{2}hx_n^2}, \quad n = 1, 2, 3, \ldots$$

with $x_0 = 2$ and $x_1 = 2 + h - \frac{1}{2}h^2$.

If instead of $x'(0) = 1$ the second initial value is $x'(0) = 0$, then we have $x'(0) = 0 \Rightarrow x_1 - x_{-1} = 0$ and putting $n = 0$ in the difference equation gives

$$\frac{x_1 - 2x_0 + x_{-1}}{h^2} - x_0 = 1 \Rightarrow x_1 = x_0 + \frac{1}{2}h^2(1 + x_0).$$

In this case the starting values for the difference equation are $x_0 = 2$ and $x_1 = 2 + \frac{3}{2}h^2$.

6.8.2 Example

Set up a spreadsheet to solve the problem of Example 6.7.2 ,

$$\frac{d^2x}{dt^2} + 2\frac{dx}{dt} + 2x = R\sin(t) \text{ with } x(0) = 0 = x'(0)$$

using the difference approximations discussed above.

The difference equation (after rearrangement) is

$$x_{n+1} = 2(1-h)x_n - \frac{1-h}{1+h}x_{n-1} + \frac{Rh^2}{1+h}\sin(t_n), \quad n = 1, 2, 3,...$$

and the initial conditions give $x_0 = 0$ and $x_1 - x_{-1} = 0$. Combining the latter with the difference equation when $n = 0$ gives $x_1 + x_{-1} = 0$ and so it follows that $x_1 = 0$. A spreadsheet for this calculation is shown in Screen 6.8.1, with $h = 0.1$ and $0 \leq t \leq 15$.

The difference between the exact solution (column C) and the difference approximation (column B) is plotted in Graph 6.8.1. Note that the percentage error quickly settles to about 0.12%, with fluctuations to larger values when x passes through zero, and the error is biggest where x changes most rapidly, as would be expected. The accuracy is comparable to that given by the modified Euler method in Exercise 6.7(1).

The difference approximation method is clearly inferior to (say) the 4th-order Runge-Kutta method for the solution of initial-value problems. An extra motive for introducing it here is that these approximations are useful when we turn to consider boundary-value problems and partial differential equations.

	A	B
1	Example	6.8.2
2	Difference	approximation
3	Driven	damped oscillator
4		
5		
6	t(n)	x(n) difference eqn
7	=0	=D2
8	=A7+D$1	=D3
9	=A8+D$1	=2*(1-D$1)*B8-(1-D$1)*B7/(1+D$1)+D$4*D$1^2*SIN(A8)/(1+D$1)

	C	D	E
1	step-size h =	=0.1	
2	x0 =	=0	
3	x1 =	=0	
4	R =	=1	
5	phi =	=ATAN(2)	
6	x(n) exact	% error	error in x
7	=D$4*(EXP(-A7)*SIN(A7+D$5)+SIN(A7-D$5))/SQRT(5)		=C7-B7
8	=D$4*(EXP(-A8)*SIN(A8+D$5)+SIN(A8-D$5))/SQRT(5)	=100*E8/C8	=C8-B8

	A	B	C	D	E
1	Example	6.8.2	step-size h =	0.1	
2	Difference	approximation	x0 =	0	
3	Driven	damped oscillator	x1 =	0	
4			R =	1	
5			phi =	1.10715	
6	t(n)	x(n) difference eqn	x(n) exact	% error	error in x
7	0	0	0		0.0E+00
8	0.1	0	0.00016	100.00	1.6E-04
9	0.2	0.00091	0.00120	24.55	3.0E-04

Screen 6.8.1 *Solution of a 2nd order ODE using difference approximations.*

Exercises 6.8

Apply the difference approximation method and the 4th-order Runge-Kutta method to the problems of (a) Example 6.7.1, (b) Example 6.8.1, and (c) Exercise 6.7(2). The latter method can serve to provide a solution (for comparison) that is very accurate, provided that the step-size is not too large. The suitability of a chosen step-size can always be checked by varying it and observing the effect on the solution.

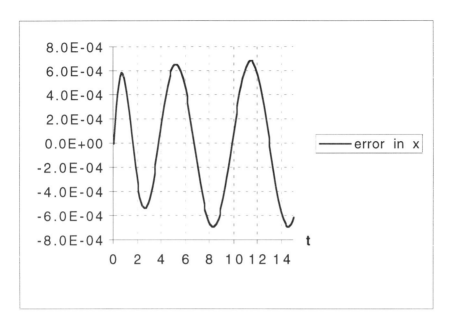

Graph 6.8.1 *Example 6.8.2: plot of difference between exact and difference approximation solutions.*

§6.9   Linear Boundary-Value Problems

Suppose we have a second-order linear ODE of the form

$$\frac{d^2y}{dx^2} + p(x)\frac{dy}{dx} + q(x)y = r(x) \qquad (6.9.1)$$

on the interval $[a,b]$ with the boundary conditions $y(a) = \alpha, y(b) = \beta$.

This is different from an initial-value problem which would specify y and its derivative both at the *same* end ($x = a$) of the interval. A physical example of a boundary-value problem would be the equation describing the deflection of a loaded beam fixed at both ends.

As before we will divide the interval $[a,b]$ into N sub-intervals, each of length $h = (b-a)/N$.

The mesh points lie at $x_k = a + kh, k = 0, 1, 2, ..., N$. The boundary points are where $k = 0$ or $k = N$, corresponding to $x = a$ or $x = b$ respectively, and the boundary conditions are $y_0 = \alpha$ and $y_N = \beta$.

Use of the central differences for the derivatives at the interior points gives, for $n = 1, 2, ..., N-1$,

$$\frac{y_{n+1} - 2y_n + y_{n-1}}{h^2} + p(x_n)\frac{y_{n+1} - y_{n-1}}{2h} + q(x_n)y_n = r(x_n)$$

$$\Rightarrow [1 - \tfrac{1}{2}hp(x_n)]y_{n-1} + [-2 + h^2 q(x_n)]y_n + [1 + \tfrac{1}{2}hp(x_n)]y_{n+1} = h^2 r(x_n)$$

Remembering that $y_0 = \alpha$ and $y_N = \beta$, and setting $n = 1, 2, 3, ..., N-1$ we get the $N-1$ equations:

$$n = 1: [-2 + h^2 q(x_1)]y_1 + [1 + \tfrac{1}{2}hp(x_1)]y_2$$
$$= h^2 r(x_1) - [1 - \tfrac{1}{2}hp(x_1)]\alpha$$

$$n = 2: [1 - \tfrac{1}{2}hp(x_2)]y_1 + [-2 + h^2 q(x_2)]y_2 + [1 + \tfrac{1}{2}hp(x_2)]y_3$$
$$= h^2 r(x_2)$$

$$\vdots$$

$$[1 - \tfrac{1}{2}hp(x_{N-2})]y_{N-3} + [-2 + h^2 q(x_{N-2})]y_{N-2} + [1 + \tfrac{1}{2}hp(x_{N-2})]y_{N-1}$$
$$= h^2 r(x_{N-2})$$

for $n = N-2$, and finally, for $n = N-1$,

$$[1 - \tfrac{1}{2}hp(x_{N-1})]y_{N-2} + [-2 + h^2 q(x_{N-1})]y_{N-1}$$
$$= h^2 r(x_{N-1}) - [1 + \tfrac{1}{2}hp(x_{N-1})]\beta$$

These equations can be put in matrix form $\mathbf{Ay} = \mathbf{b}$ with $\mathbf{A}$ being the $(N-1) \times (N-1)$ tri-diagonal matrix with elements a_{ij} given by

$$a_{ij} = 0, \qquad\qquad j < i-1 \text{ or } j > i+1$$
$$a_{j\,j-1} = 1 - \tfrac{1}{2}hp(x_j), \quad j = 2, 3,..., N-1$$
$$a_{j\,j} = -2 + h^2 q(x_j), \quad j = 1, 2, 3,..., N-1 \qquad (6.9.2)$$
$$a_{j\,j+1} = 1 + \tfrac{1}{2}hp(x_j), \quad j = 1, 2, 3,..., N-2$$

and **y** and **b** are $(N-1) \times 1$ column vectors with

$$b_1 = h^2 r(x_1) - [1 - \tfrac{1}{2}hp(x_1)]\alpha$$
$$b_j = h^2 r(x_j), \qquad j = 2, 3,..., N-2 \qquad (6.9.3)$$
$$b_{N-1} = h^2 r(x_{N-1}) - [1 + \tfrac{1}{2}hp(x_{N-1})]\beta.$$

Sufficient conditions for the linear boundary value problem (6.9.1) and the tridiagonal system $\mathbf{Ay} = \mathbf{b}$ to have a unique solution are described in the following theorem:

6.9.1 Theorem

If the functions $p(x)$, $q(x)$ and $r(x)$ are continuous and $q(x) \le 0$ on $[a,b]$ then

(i) the linear boundary value problem (6.9.1) has a unique solution, and

(ii) the tridiagonal system $\mathbf{Ay} = \mathbf{b}$ has a unique solution, provided that $h < 2/M$, where M is the maximum value of $p(x)$ on $[a,b]$.

6.9.2 Example

Apply the method outlined above to the boundary-value problem

$$\frac{d^2y}{dx^2} + \frac{1}{x}\frac{dy}{dx} - \frac{1}{x^2}y = \frac{1}{x^2}\cos(\ln(x))$$

on the interval $[1,2]$ *with boundary conditions* $y(1)=2$, $y(2)=0$. *The exact solution is*

$$y = \frac{(2\cos(\ln(2))-5)x}{6} + \frac{10-\cos(\ln(2))}{3x} - \frac{\cos(\ln(x))}{2}.$$

In this example $p(x)=1/x$ and $q(x)=-1/x^2$ and hence the matrix elements (6.9.2) are:

Sub-diagonal: $a_{j\,j-1} = 1 - \dfrac{h}{2x_j}$, $\quad j=2,3,...,N-1$

Diagonal: $a_{jj} = -2 - \dfrac{h^2}{x_j^2}$, $\quad j=1,2,3,...,N-1$

Super-diagonal: $a_{j\,j+1} = 1 + \dfrac{h}{2x_j}$, $\quad j=1,2,3,...,N-2$

while the right-hand side terms (6.9.3) are:

$$b_1 = \frac{h^2\cos(\ln(x_1))}{x_1^2} - 2 + \frac{h}{x_1}, \quad b_{N-1} = \frac{h^2\cos(\ln(x_{N-1}))}{x_{N-1}^2},$$

and

$$b_j = \frac{h^2\cos(\ln(x_j))}{x_j^2}, \quad j=2,3,...,N-2.$$

Theorem 6.9.1 ensures a unique solution for $h<2$, since $|p(x)|\le 1$ on the interval $[1,2]$.

A spreadsheet for the case $N = 10, h = 0.1$ is shown in screen 6.9.1, with exact and numerical solutions and the percentage error plotted in Graph 6.9.1. The spreadsheet is laid out exactly as in Example 3.4.4 (Crout's method for solving a tri-diagonal system of equations). The chief difference from that case is that here the matrix elements and the right-hand side terms are functions of the distance x along the interval $[1,2]$.

	A	B	C	D
1	Example 6.9.2			
2	Linear BVP			
3		tri-	diagonal	equations
4		x(0)=a=	=1	x(N)=b=
5		y(a) =	=2	y(b) =
6		N =	=10	h=(b-a)/N=
7			Matrix elements	
8	i	x(i)	sub-diags	diags
9	=0	=C$4		
10	=1	=B9+E$6		=-2-G$6/B10^2
11	=A10+1	=B10+E$6	=1-0.5*E$6/B11	=-2-G$6/B11^2

	A	B	C	D
17	=A16+1	=B16+E$6	=1-0.5*E$6/B17	=-2-G$6/B17^2
18	=A17+1	=B17+E$6	=1-0.5*E$6/B18	=-2-G$6/B18^2
19	=A18+1	=B18+E$6		

	E	F	G
4	=2		
5	=0		
6	=(E4-C4)/C6	h^2 =	=E6^2
7			
8	super-diags	RHS	p(i)
9			
10	=1+0.5*E$6/B10	=G$6*COS(LN(B10))/B10^2-(2-E6/B10)	=D10
11	=1+0.5*E$6/B11	=G$6*COS(LN(B11))/B11^2	=D11-C11*H10

	E	F	G
17	=1+0.5*E$6/B17	=G$6*COS(LN(B17))/B17^2	=D17-C17*H16
18		=G$6*COS(LN(B18))/B18^2	=D18-C18*H17

Screen 6.9.1 *Solving a linear boundary-value problem using difference approximations.*

223

	H	I	J	K
4			c1 =	=(COS(LN(2))-2.5)/3
5			c2 =	=2.5-K4
6				
7				(num)
8	q(i)	z(i)	x(i)	y(i)
9			=C$4	=C5
10	=E10/G10	=F10/G10	=J9+E$6	=I10-K11*H10
11	=E11/G11	=(F11-C11*I10)/G11	=J10+E$6	=I11-K12*H11

	H	I	J	K
17	=E17/G17	=(F17-C17*I16)/G17	=J16+E$6	=I17-K18*H17
18		=(F18-C18*I17)/G18	=J17+E$6	=I18
19			=J18+E$6	=E5

	L	M	N
7	(exact)		
8	y(i)	% error	error
9	=K$4*J9+K$5/J9-0.5*COS(LN(J9))	=0	
10	=K$4*J10+K$5/J10-0.5*COS(LN(J10))	=100*N10/L10	=L10-K10

	L	M	N
18	=K$4*J18+K$5/J18-0.5*COS(LN(J18))	=100*N18/L18	=L18-K18
19	=K$4*J19+K$5/J19-0.5*COS(LN(J19))	=0	

	A	B	C	D	E	F	G
1	Example 6.9.2						
2	Linear BVP						
3		tri-	diagonal	equations			
4	x(0)=a=		1	x(N)=b=	2		
5	y(a) =		2	y(b) =	0		
6		N =	10	=(b-a)/N=	0.1	h^2 =	0.01
7			Matrix elements				
8	i	x(i)	sub-diags	diags	super-diags	RHS	p(i)
9	0	1					
10	1	1.1		-2.0083	1.0455	-1.901	-2.008
11	2	1.2	0.9583	-2.0069	1.0417	0.007	-1.508

	A	B	C	D	E	F	G
17	8	1.8	0.9722	-2.0031	1.0278	0.003	-1.135
18	9	1.9	0.9737	-2.0028		0.002	-1.121
19	10	2					

Screen 6.9.1 (cont)

	H	I	J	K	L	M	N
4			c1 =	-0.5769204			
5			c2 =	3.0769204			
6							
7				(num)	(exact)		
8	q(i)	z(i)	x(i)	y(i)	y(i)	% error	error
9			1	2.00	2.00	0	
10	-0.521	0.947	1.1	1.67	1.66	-0.02702	-0.00045
11	-0.691	0.597	1.2	1.38	1.38	-0.04843	-0.00067

	H	I	J	K	L	M	N
17	-0.906	0.144	1.8	0.26	0.25	-0.11890	-0.00030
18		0.123	1.9	0.12	0.12	-0.12561	-0.00015
19			2	0.00	0.00	0	

Screen 6.9.1 (cont)

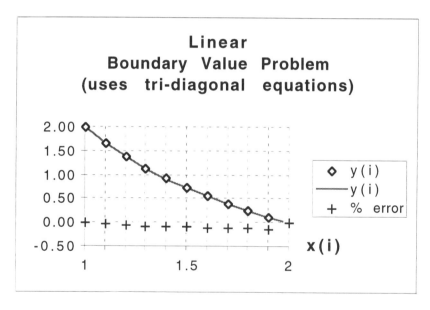

Graph 6.9.1

Exercises 6.9.1

Solve the following linear boundary-value problems using the method described above. Begin with $N = 10$ and later extend your spreadsheets to larger values of N.

(a)

$$\frac{d^2 y}{dx^2} + 4\sin(x)\frac{dy}{dx} - 4\cos(x)y = -\sin(x)$$

on $[0, \frac{\pi}{2}]$ with $y(0) = 0$, $y(\frac{\pi}{2}) = 1$,

(b)

$$\frac{d^2 y}{dx^2} + 3\frac{dy}{dx} + 2y = 2x + 5$$

on $[0,1]$ with $y(0) = 1$, $y(1) = 3$,

The exact solutions are

(a) $y = \sin(x)$ and (b) $y = \dfrac{e^2}{e-1}(e^{-x} - e^{-2x}) + x + 1$.

6.9.3 The Shooting Method for Boundary Value Problems

Another way of solving boundary-value problems is the so-called shooting method. We will consider its application to linear equations like those discussed above.

The original problem (6.9.1) can be re-stated in terms of the solution of two initial-value problems on the interval $[a,b]$. Suppose $y_1(x)$ and $y_2(x)$ are the solutions of the initial-value problems

$$\frac{d^2 y_1}{dx^2} + p(x)\frac{dy_1}{dx} + q(x)y_1 = r(x) \text{ with } y_1(a) = \alpha, \quad y_1'(a) = 0 \quad (6.9.4)$$

and

$$\frac{d^2 y_2}{dx^2} + p(x)\frac{dy_2}{dx} + q(x)y_2 = 0 \text{ with } y_2(a) = 0, \quad y_2'(a) = 1 \quad (6.9.5)$$

respectively. Then the function

$$y(x) = y_1(x) + \frac{\beta - y_1(b)}{y_2(b)} y_2(x)$$

is the unique solution to the linear boundary-value problem (6.9.1), provided $y_2(b) \neq 0$. We can use any of the numerical methods noted earlier in this chapter to solve the initial-value problems (6.9.4) and (6.9.5).

6.9.4   Example

Apply the shooting method to the problem of Example 6.9.2,

$$\frac{d^2 y}{dx^2} + \frac{1}{x}\frac{dy}{dx} - \frac{1}{x^2}y = \frac{1}{x^2}\cos(\ln(x)) \text{ on } [1,2] \text{ with } y(1) = 2, \ y(2) = 0$$

using the modified Euler method to solve the two initial-value problems.

See Screen 6.9.2 and Graph 6.9.2. Comparison with Graph 6.9.1 shows the error to be roughly comparable for these two methods.

For this example, the two initial-value problems (expressed as pairs of first-order equations) are:

$$\frac{dy_1}{dx} = z_1, \qquad\qquad\qquad y_1(1) = 2$$

$$\frac{dz_1}{dx} = -\frac{1}{x}z_1 + \frac{1}{x^2}y_1 + \frac{1}{x^2}\cos(\ln(x)), \quad z_1(1) = 0$$

and

$$\frac{dy_2}{dx} = z_2, \qquad\qquad y_2(1) = 0$$

$$\frac{dz_2}{dx} = -\frac{1}{x}z_2 + \frac{1}{x^2}y_2, \quad z_2(1) = 1$$

For the first initial value problem (tabulated in columns B to H), the use of equations (6.6.2) requires the functions

$$k_1(n) = f(x_n, y_{1n}, z_{1n}) = z_{1n}$$

$$l_1(n) = g(x_n, y_{1n}, z_{1n}) = -\frac{z_{1n}}{x_n} + \frac{y_{1n} + \cos(\ln(x_n))}{x_n^2}$$

$$k_2(n) = f(x_{n+1}, y_{1n} + hk_1(n), z_{1n} + hl_1(n)) = z_{1n} + hl_1(n)$$

$$l_2(n) = g(x_{n+1}, y_{1n} + hk_1(n), z_{1n} + hl_1(n))$$

$$= -\frac{z_{1n} + hl_1(n)}{x_n + h} + \frac{y_{1n} + hk_1(n) + \cos(\ln(x_n + h))}{(x_n + h)^2}$$

The second initial value problem (tabulated in columns B and I to N) requires

$$k_1(n) = f(x_n, y_{2n}, z_{2n}) = z_{2n}$$

$$l_1(n) = g(x_n, y_{2n}, z_{2n}) = -\frac{z_{2n}}{x_n} + \frac{y_{2n}}{x_n^2}$$

$$k_2(n) = f(x_{n+1}, y_{2n} + hk_1(n), z_{2n} + hl_1(n)) = z_{2n} + hl_1(n)$$

$$l_2(n) = g(x_{n+1}, y_{2n} + hk_1(n), z_{2n} + hl_1(n)) = -\frac{z_{2n} + hl_1(n)}{x_n + h} + \frac{y_{2n} + hk_1(n)}{(x_n + h)^2}$$

The boundary value problem numerical solution lies in column O, with the exact solution in column P and the percentage error in column Q.

	A	B	C
1	Example 6.9.4		
2	Linear BVP		
3			Shooting
4		x(0)=a=	=1
5		y(a) =	=2
6		N =	=10
7			First IVP
8	i	x(i)	y1(i)
9	=0	=C$4	=C5
10	=A9+1	=B9+E$6	=C9+E$6*(E9+F9)/2

	D	E	F
3	Method		
4	x(N)=b=	=2	
5	y(b) =	=0	
6	h=(b-a)/N=	=(E4-C4)/C6	
7			
8	z1(i)	k1(i)	k2(i)
9	=0	=D9	=D9+E$6*G9
10	=D9+E$6*(G9+H9)/2	=D10	=D10+E$6*G10

	G
8	l1(i)
9	=(C9+COS(LN(B9)))/B9^2-D9/B9
10	=(C10+COS(LN(B10)))/B10^2-D10/B10

	H
8	l2(i)
9	=(C9+E$6*E9+COS(LN(B9+E$6)))/(B9+E$6)^2-(D9+E$6*G9)/(B9+E$6)

	I	J	K	L
7	Second IVP			
8	y2(i)	z2(i)	k1(i)	k2(i)
9	=0	=1	=J9	=J9+E$6*M9
10	=I9+E$6*(K9+L9)/2	=J9+E$6*(M9+N9)/2	=J10	=J10+E$6*M10

	M	N
7		
8	l1(i)	l2(i)
9	=I9/B9^2-J9/B9	=(I9+E$6*K9)/(B9+E$6)^2-(J9+E$6*M9)/(B9+E$6)

	O	P
7	BVP	
8	y(i) num	y(i) exact
9	=C9+(E$5-C$19)*I9/I$19	=P$5*B9+P$6/B9-COS(LN(B9))/2

Screen 6.9.2 *Solution of a linear boundary value problem by the shooting method, and using the modified Euler method to solve the initial value problems.*

	O	P
5	c1 =	=(COS(LN(2))-2.5)/3
6	c2 =	=2.5-P5

	Q	R
8	**% error**	error
9		
10	=100*R10/P10	=P10-O10

	A	B	C	D	E	F	G	H
1	Example 6.9.4							
2	Linear BVP							
3			Shooting	Method				
4	x(0)=a=	1		x(N)=b=	2			
5	y(a) =	2		y(b) =	0			
6		N =	10	h=(b-a)/N=	0.1			
7			First IYP					
8	i	x(i)	y1(i)	z1(i)	k1(i)	k2(i)	11(i)	12(i)
9	0	1	2.0000	0.0000	0.000	0.300	3.000	2.203

	A	B	C	D	E	F	G	H
18	9	1.9	2.6374	1.0622	1.062	1.102	0.393	0.327
19	10	2	2.7456	1.0983	1.098	1.131	0.330	0.276

	I	J	K	L	M	N
7	Second IYP					
8	y2(i)	z2(i)	k1(i)	k2(i)	11(i)	12(i)
9	0.0000	1.0000	1.000	0.900	-1.000	-0.736

	I	J	K	L	M	N
18	0.6852	0.6382	0.638	0.624	-0.146	-0.125
19	0.7483	0.6247	0.625	0.612	-0.125	-0.108

	O	P	Q	R
5	c1 =	-0.57692037		
6	c2 =	3.076920366		
7	BVP			
8	y(i) num	y(i) exact	% error	error
9	2.00000	2.00000		
10	1.66641	1.66486	-0.093	-0.00155

	O	P	Q	R
18	0.12331	0.12279	-0.421	-0.00052
19	0.00000	0.00000		

Screen 6.9.2 (continued)

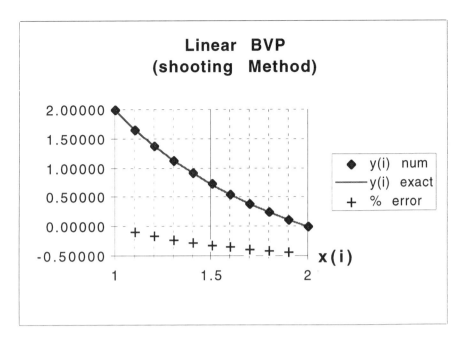

Graph 6.9.2

Exercises 6.9.2

(1) Solve the problems of Exercises 6.9.1 using the shooting method. Try both modified Euler and Runge-Kutta methods when solving the initial-value problems that arise.

(2) Solve Example 6.9.2 using the shooting method with Runge-Kutta.

6.9.5 Increasing Accuracy using the Richardson Extrapolation

One way to get greater accuracy with the shooting method is to use a more accurate method to solve the initial-value problems, for example the Runge-Kutta method. Another is to apply extrapolation to the results of the less accurate method.

In §5.6 Richardson's extrapolation process was used to improve the accuracy of numerical integration with the trapezoidal rule. Given the connection noted in §6.4 between the trapezoidal rule and the modified Euler method, it is reasonable to expect that the same idea can be applied to increase the accuracy of results derived from the use of the modified Euler method. Furthermore, the central difference approximations (6.8.1) for first and second derivatives also have the form (5.6.2), thus making Richardson's extrapolation also applicable to solutions obtained with the finite difference method.

The outline of the first stage of the extrapolation process is as follows: First, Solve the BVP with step-sizes h and $\frac{1}{2}h$. Let $y_k(h)$ and $y_{2k}(\frac{1}{2}h)$ denote the respective approximate solutions at the point where $x = kh = 2k \times (\frac{1}{2}h)$, and let $y(kh)$ be the exact value. So if $y_k(h) = y(kh) + Ah^2 + O(h^4)$, then $y_{2k}(\frac{1}{2}h) = y(kh) + \frac{1}{4}Ah^2 + O(h^4)$.

The first extrapolation is then given by elimination of the $O(h^2)$ term:

$$y(kh) \approx \frac{4y_{2k}(\frac{1}{2}h) - y_k(h)}{3} \qquad (6.9.6)$$

This result has truncation error which is $O(h^4)$. Greater accuracy can be reached by further extrapolation as described in §5.6.

6.9.6   Example

Apply one step of the Richardson extrapolation to the problem of Example 6.9.2.

The spreadsheet for Example 6.9.2 can be extended reasonably easily so that it also includes the same calculation with half the step-size. Just make a copy of the original procedure a few rows below it, and amend this copy by careful use of the FILL command so that it caters for a halved step-size. The first extrapolation can then be applied. Screen 6.9.3 shows only the extrapolation part of the formulas and results displayed in such a spreadsheet.

The first "copy" of the numerical BVP solution procedure (for step-size $h = 0.1$) has its results stored in cells K9 to K19 (as in Screen 6.9.1), and the second set of results (for step-size $h = 0.05$) are stored in cells K29 to K49. The extrapolation formulas corresponding to eq. (6.9.6) are in cells P9 to P19.

The errors for the case $h = 0.1$ are less than about 7.5×10^{-4}, for the case $h = 0.05$ they are less than about 1.9×10^{-4} (showing the expected reduction by a factor of about 4), and the extrapolated results show errors of less than about 1.4×10^{-6}. This significant improvement has been gained with much less extra computation than that needed to reach the same accuracy simply by using a sufficient reduction of the step-size.

	0	P	Q	R
8	x(i)	y(i) (extrap)	% error (extrap)	error
9	=C$4	=(4*K29-K9)/3		
10	=O9+E$6	=(4*K31-K10)/3	=100*R10/L10	=L10-P10

	0	P	Q	R
18	=O17+E$6	=(4*K47-K18)/3	=100*R18/L18	=L18-P18
19	=O18+E$6	=(4*K49-K19)/3		

	0	P	Q	R
8	x(i)	y(i) (extrap)	% error (extrap)	error
9	1	2.00000000		
10	1.1	1.66485817	-5.69E-05	-9.47E-07

	0	P	Q	R
18	1.9	0.12278967	-1.81E-04	-2.23E-07
19	2	0.00000000		

Screen 6.9.3 Richardson extrapolation is used here to obtain a significant increase in the accuracy of the solution of a linear BVP, based on the use of difference approximations and Crout's algorithm.

Exercises 6.9.3

(1) Perform the above-mentioned extension of the spreadsheet of Example 6.9.2.

(2) Apply extrapolation to the shooting method solution of Example 6.9.4.

(3) Repeat the exercises (1) and (2) above for the problems of Exercises 6.9.1

(4) Use extrapolation of the modified Euler method to solve the problems of Exercise 6.2(2). Compare the accuracy with that found in Exercises 6.3 where the same problems were solved using the 4th-order Runge-Kutta method.

§6.10* Stiff Differential Equations

Systems of differential equations arise in diverse areas of applied mathematics, for example damped spring systems in mechanics, control systems, and in chemical kinetics. A common feature is the coexistence of long-term "steady state" behaviour with relatively rapid transient behaviour. It is the latter that constitutes the "stiff" part of the evolution of the system under study.

A single differential equation can exhibit both transient and slow behaviour: The equation $x' = \lambda(x - F(t)) + F'(t)$ has solution $x(t) = (x_0 - F(0))e^{\lambda t} + F(t)$, and if $\lambda << 0$ this behaves as $F(t)$ for all but very small values of t, so the steady state part of the solution is $F(t)$, with the first term being the transient part.

Accurate integration of the transient behaviour requires relatively small step-size, while the long-term behaviour can be adequately handled with bigger steps. The method used must remain stable when the step-size is increased as required by the problem.

As noted in §6.5, each method has its own characteristic "zone of stability", giving the range of step-sizes for which a stable solution is obtained for the **test problem** $x' = \lambda x$, $\lambda < 0$. Although most systems of interest are non-linear, it suffices to study linear systems in order to begin to explore the stability of the various numerical methods, because the local behaviour is that of the linearized version of a non-linear system.

6.10.1 Example

The linear two-variable system

$$u' = \frac{du}{dt} = 197u + 297v, \quad u(0) = 2$$
$$v' = \frac{dv}{dt} = -198u - 298v, \quad v(0) = -1$$

has solution

$$u(t) = 3e^{-t} - e^{-100t}$$
$$v(t) = -2e^{-t} + e^{-100t}$$

Each variable has a fast and a slow component, governed by e^{-100t} and e^{-t} respectively. Since this particular system of differential equations is linear with non-singular matrix it can be decoupled with the transformation

$$\begin{bmatrix} x \\ y \end{bmatrix} = \begin{bmatrix} 1 & 1 \\ 2 & 3 \end{bmatrix} \begin{bmatrix} u \\ v \end{bmatrix}$$

to give

$$x' = -x, \qquad x(0) = 1$$
$$y' = -100y, \quad y(0) = 1$$

These two equations can be solved separately, so that the restriction to smaller h required for stability for the y equation does not affect our ability to solve the x equation with sufficient accuracy using a larger value of h.

In general, however, the system of equations being studied will be nonlinear and possibly non-autonomous, and a decoupling of the variables will be difficult or impossible. The problem then is to devise a numerical method that copes adequately with the rapid transients (indicating a need for small h) without requiring a prohibitive number of steps (and consequent higher rounding error) to yield the slower parts of the solution.

Screen 6.10.1 shows a spreadsheet in which the pair of differential equations for u and v given above are solved using the modified Euler method, with a **change in step-size** at $t = Nh_1$, after N iterations with the initial step-size h_1, to $h_2 = Mh_1$.

The value of h_1 is chosen so that $D = |F(\lambda_1 h_1) - \exp(\lambda_1 h_1)| \leq 10^{-4}$ (say), where $\lambda_1 = -100$ and $F(\lambda h) = 1 + \lambda h + \frac{1}{2}(\lambda h)^2$. N is determined by the requirement that the transient term $\exp(-100t)$ should be negligible in comparison with the steady state part, $\exp(-t)$. It is important to note that the stability requirement $h_1 < 0.02$ -which arises from $h|\lambda| < 2$ with $\lambda = \lambda_1 = -100$ - must still be satisfied in the region where the terms corresponding to $\lambda_2 = -1$ are dominating the solution.

The initial step-size h_1 and the multiplier M giving the changed value $h_2 = Mh_1$ are stored in cells D1 and F2. The switch from h_1 to h_2 is effected in columns D, G and H by means of an IF function which returns the value h_1 or h_2 according as $n < N$ or not. Here, n counts the number of steps calculated, with the switch occurring at N steps.

The ratio of the transient to the steady state contributions at the switch point $t = Nh_1$ is displayed in cell F4. Graph 6.10.1 shows the solutions and Graph 6.10.2 the magnitude of the global error $\|\mathbf{g}_n\|$, where

$$\|\mathbf{g}_n\| = \sqrt{(u_n - u(t_n))^2 + (v_n - v(t_n))^2} \ .$$

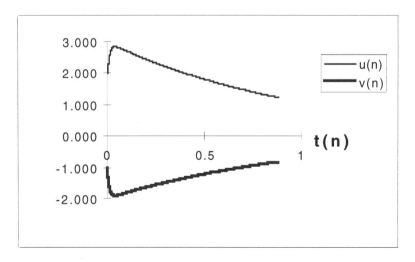

Graph 6.10.1 *u and v vs t when step-size changes from 0.0008 to 0.004 at t = 0.08*

	A	B
1	Example	
2	6.10.1	
3		
4		
5		
6	n	t(n)
7	=0	=0
8	=A7+1	=B7+IF(A7<D$3,D$1,D$2)

	C
1	h1=
2	h2=M*h1=
3	N=
4	h1->h2 at t=
5	D =
6	u(n)
7	=2
8	=C7+0.5*IF(A7<D$3,D$1,D$2)*(E7+G7)

	D
1	=0.0008
2	=F2*D1
3	=100
4	=D3*D1
5	=ABS(1-100*D1+5000*D1^2-EXP(-100*D1))
6	v(n)
7	=-1
8	=D7+0.5*IF(A7<D$3,D$1,D$2)*(F7+H7)

	E	F
2	M=	=5
3		
4	exp(-99t(N))=	=EXP(-99*D4)
5		
6	k1(n)	l1(n)
7	=197*C7+297*D7	=-198*C7-298*D7

	G
6	k2(n)
7	=197*(C7+IF(A7<D$3,D$1,D$2)*E7)+297*(D7+IF(A7<D$3,D$1,D$2)*F7)

	H
6	l2(n)
7	=-198*(C7+IF(A7<D$3,D$1,D$2)*E7)-298*(D7+IF(A7<D$3,D$1,D$2)*F7)

Screen 6.10.1 *Numerical solution of a stiff pair of differential equations using the modified Euler method. The step-size can be varied after N steps with the initial step-size.*

	I	J
6	exact u(n)	exact v(n)
7	=3*EXP(-B7)-EXP(-100*B7)	=-2*EXP(-B7)+EXP(-100*B7)

	K	L	M
6	error in u	error in v	|global error|
7	=C7-I7	=D7-J7	=SQRT((C7-I7)^2+(D7-J7)^2)

	A	B	C	D	E	F	G	H
1	Example		h1=	0.0008				
2	6.10.1	h2=M*h1=		0.0040	M=	5		
3			N=	100				
4		h1->h2 at t=		0.0800	exp(-99t(N))=	0.0004		
5			D =	0.0001				
6								
7	n	t(n)	u(n)	v(n)	k1(n)	l1(n)	k2(n)	l2(n)
8	0	0	2.000	-1.000	97.00	-98.00	89.00	-90.00
9	1	0.0008	2.074	-1.075	89.32	-90.32	81.94	-82.94

	I	J	K	L	M
7	exact u(n)	exact v(n)	error in u	error in v	|global error|
8	2.0000	-1.0000	0.00E+00	0.00E+00	0
9	2.0745	-1.0753	-8.37E-05	8.37E-05	0.000118304

Screen 6.10.1 (*continued*)

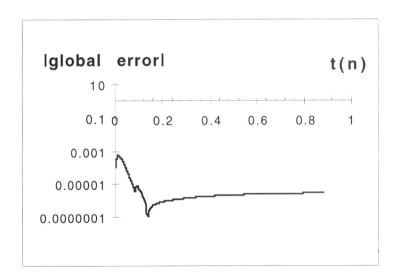

Graph 6.10.2 *Global error when step-size changes from 0.0008 to 0.004 at*
t = 0.08

There is a rise in the value of $\|g_n\|$ when the step-size is increased, due to the error that already exists at that point.

Some comparisons are shown below, listing the step-sizes, total number N_T of steps to reach $t = T = 0.88$, and an upper bound G for the size of the global error, with the change in step-size (if any) occurring at $t = 0.08$.

h_1	h_2	N_T	G
0.004	0.004	220	0.019
0.0008	0.0008	1100	0.0006
0.0008	0.004	300	0.0006
0.0008	0.016	150	0.0006

The advantage in terms of reducing the number of steps without loss of accuracy is clearly demonstrated. See Exercises 6.10 for further exploration of this and other problems.

The decay constants $\lambda_1 = -100$ and $\lambda_2 = -1$ that arise in Example 6.10.1 are the eigenvalues of the coefficient matrix $\mathbf{M}$ for the linear ODE's:

$$\begin{bmatrix} u' \\ v' \end{bmatrix} = \mathbf{M} \begin{bmatrix} u \\ v \end{bmatrix}, \text{ where } \mathbf{M} = \begin{bmatrix} 197 & 297 \\ -198 & -298 \end{bmatrix}.$$

A measure of the **stiffness** S of the system of equations is the ratio of the magnitudes of the largest and smallest eigenvalues. Thus for Example 6.10.1 the stiffness has the not particularly high value

$$S = \frac{|-100|}{|-1|} = 100$$

The behaviour of a system of nonlinear equations over each small step is governed by the Jacobi matrix $\mathbf{J}$ at the current point on the trajectory, because the coefficients of the linearized form of the equations are the elements of $\mathbf{J}$. As the system evolves the elements of $\mathbf{J}$ will vary, and so the eigenvalues of $\mathbf{J}$ will vary, unlike the case of linear equations where they are constant. The stability constraint on h imposed by the most negative eigenvalue may therefore also vary for nonlinear equations. An example of this type can be found in Schwarz [1989].

6.10.2 Implicit Methods

In the preceding examples and exercises the use of explicit methods incurs a restriction on the available choice of step-size because of stability requirements. When very stiff equations are involved, the greater freedom afforded by implicit methods becomes essential. One such method already mentioned is the implicit trapezoidal method mentioned in §6.5, $x_{n+1} = x_n + \frac{1}{2}h[f(x_{n+1}) + f(x_n)]$, for which the stability condition is

$$\left| \frac{1 + \frac{1}{2}\lambda h}{1 - \frac{1}{2}\lambda h} \right| < 1$$

With $\lambda < 0$ this is satisfied for all $h > 0$.

Another is the implicit 2nd order Runge-Kutta method,

$$x_{n+1} = x_n + hk_1(n), \quad k_1(n) = f(t_n + \frac{1}{2}h, x_n + \frac{1}{2}hk_1(n)) \qquad (6.10.1)$$

When the latter method is applied to the standard equation $x' = \lambda x$ the stability condition that results is the same as for the implicit trapezoidal method.

For a system of two equations the implicit 2nd order Runge-Kutta method is

$$x_{n+1} = x_n + hk_1(n), \quad y_{n+1} = y_n + hl_1(n), \text{ where}$$
$$k_1(n) = f(t_n + \frac{1}{2}h, x_n + \frac{1}{2}hk_1(n), y_n + \frac{1}{2}hl_1(n)) \qquad (6.10.2)$$
$$l_1(n) = g(t_n + \frac{1}{2}h, x_n + \frac{1}{2}hk_1(n), y_n + \frac{1}{2}hl_1(n))$$

The use of implicit methods on a spreadsheet is not very convenient for nonlinear equations. For a pair of linear equations of the form

$$x' = f(t, x, y) = ax + by$$
$$y' = g(t, x, y) = cx + dy$$

the implicit 2nd order Runge-Kutta method becomes the explicit equations

$$x_{n+1} = \frac{[4+2(a-d)h-(ad-bc)h^2]x_n + 4bhy_n}{D}$$

$$y_{n+1} = \frac{4chx_n + [4+2(d-a)h-(ad-bc)h^2]y_n}{D}$$

(6.10.3)

where $ad-bc$ and $D = 4-2(a+d)+(ad-bc)h^2$ are both assumed to be non-zero.

6.10.3   Example

Apply the pair of equations (6.10.3) to the problem of Example 6.10.1, to allow comparison with the explicit modified Euler method employed there.

Some results are tabulated below for the case where the change in step-size is at $t = 0.08$. The details of the spreadsheet are left as an exercise (see Exercise 6.10(5)(a) below).

h_1	h_2	N_T	G
0.004	0.004	220	0.007
0.0008	0.0008	1100	0.0003
0.0008	0.004	300	0.0003
0.0008	0.016	150	0.0003
0.0008	0.032	125	0.0003

It should be noted that the modified Euler method is unstable when used with the value $h_2 = 0.032 > 0.02$ and thus an advantage of the implicit method has been clearly demonstrated. Graph 6.10.4 shows the global error for the last case tabulated.

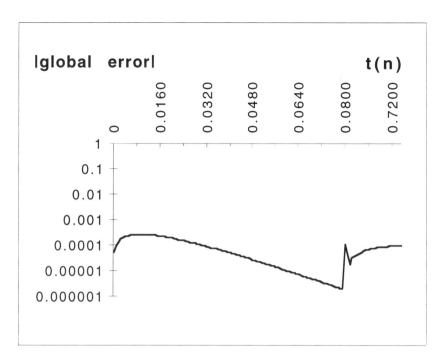

Graph 6.10.4 *Global error when the implicit 2nd order Runge Kutta method is used, with step sizes 0.0008 and 0.032, changing at t = 0.08.*

Exercises 6.10

(1) Experiment with the spreadsheet of Example 6.10.1. Vary the initial step-size, the point at which the step-size changes, and the magnitude of the change. The effects are probably best seen in a plot like Graph 6.10.3.

(2) Construct a spreadsheet for the problem of Example 6.10.1 using the 4th-order Runge-Kutta method, and explore its behaviour in a manner similar to (1) above.

(3) Apply the modified Euler and 4th-order Runge-Kutta methods to the very stiff linear system

$$u' = \frac{du}{dt} = 10u - 10^3 v, \quad u(0) = 1$$
$$v' = \frac{dv}{dt} = 101u - 10^4 v, \quad v(0) = -1$$

for which the eigenvalues are $\lambda_1 \approx -9990$ and $\lambda_2 \approx -0.1$, and the stiffness is about 10^5. Introduce a change in step-size at an appropriate point, choosing the initial step-size so that $|F(\lambda h_1) - \exp(\lambda h_1)| < 10^{-4}$, to begin with.

(4) Derive the stability condition for the implicit 2nd-order Runge-Kutta method.

(5) Apply the implicit 2nd-order Runge-Kutta method to the problems of (a) Example 6.10.1 and (b) Exercise 6.10(3), and experiment with the step-sizes and the position of the change in step-size.

§6.11* Automatic Step-size Control

It would obviously be very useful to be able to estimate the local truncation error d_k at the kth step, since that would afford some control of the global error. Most simply, if $|d_k|$ exceeds some pre-determined amount then reduce the step-size used for the next step. On the other hand, if $|d_k|$ is too small then the step size could be increased. Such a process would help to minimize the number of steps needed to get a specified accuracy.

Of course there is a price to pay, namely the extra calculations needed to get an estimate of $|d_k|$. One approach is to use two Runge-Kutta methods, one of order n, the other of order $n+1$, arranged so that they share most of the function evaluations that are needed.

One such pair is the (2nd-order) improved polygonal Euler method,

$$x_{k+1} = x_k + hk_2(n),$$

where

$$k_1(n) = f(t_n, x_n)$$
$$k_2(n) = f(t_n + \tfrac{1}{2}h, x_n + \tfrac{1}{2}hk_1(n))$$

plus Kutta's 3rd-order method,

$$x_{k+1} = x_k + \tfrac{1}{6}h(k_1(n) + 4k_2(n) + k_3(n)),$$

for which $k_1(n)$ and $k_2(n)$ are as above, and

$$k_3(n) = f(t_n + h, x_n - hk_1(n) + 2hk_2(n)).$$

For this combination it can be shown that (for the former method)

$$d_{k+1} \approx \tfrac{1}{6}h(k_1 - 2k_2 + k_3) + O(h^4) \qquad (6.11.1)$$

6.11.1 Example

Apply the method outlined above to the initial-value problem

$$\frac{dx}{dt} = -2tx, \quad x(0) = 1$$

which has exact solution $x = \exp(-t^2)$.

In Screen 6.11.1 this problem is solved numerically using the improved polygonal Euler method, with step-size control derived from the expression (6.11.1) for the local truncation error.

	A	B	C	D
1	Example		initial h =	=0.04
2	6.11.1		dmax =	=0.000001
3				
4	n	t(n)	x(n)	k1(n)
5	=0	=0	=1	=-2*B5*C5
6	=A5+1	=B5+H5	=C5+H5*E5	=-2*B6*C6

	E
4	k2(n)
5	=-2*(B5+H5/2)*(C5+H5*D5/2)

	F	G
4	k3(n)	d(n)
5	=-2*(B5+H5)*(C5-H5*D5+2*H5*E5)	
6	=-2*(B6+H6)*(C6-H6*D6+2*H6*E6)	=H5*(D5-2*E5+F5)/6

	H
4	h
5	=D1
6	=IF(ABS(G6)<D$2/10,2*H5,IF(ABS(G6)>10*D$2,H5/2,H5))

	I	J
4	exact x	error
5	=EXP(-(B5^2))	=C5-I5

	A	B	C	D	E	F
1	Example 6.11.1		initial h =	0.04		
2			dmax =	0.000001		
3						
4	n	t(n)	x(n)	k1(n)	k2(n)	k3(n)
5	0	0	1.00	0.0000	-0.0400	-0.0797
6	1	0.04	1.00	-0.0799	-0.1196	-0.1587

	G	H	I	J
4	d(n)	h	exact x	error
5		0.04	1	0
6	1.71E-06	0.04	0.9984	-1.3E-06

Screen 6.11.1 *Automatic step-size control for the 2nd-order improved plygonal Euler method using Kutta's 3rd-order method to estimate the local truncation error.*

The spreadsheet also includes the solution using the improved polygonal method without step-size control (to provide data for Graph 6.11.2), but these formulas are not shown in Screen 6.11.1. The reader should add this part as an exercise.

Graph 6.11.1 shows the step-size variations that result when $d_{max} = 10^{-6}$ and initially $h = 0.04$. Graph 6.11.2 shows the actual error with and without step-size control, the latter being done with step-size fixed at the initial value. Graph 6.11.3 shows the solution.

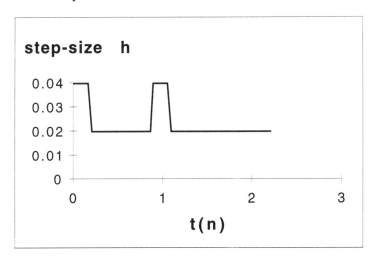

Graph 6.11.1 *The variation of the step-size h when the initial value is 0.04.*

The control is effected as follows: The value of the local truncation error d_n at each step is calculated in column G. In column H the value of h is either doubled or halved according as to whether $|d_n| < d_{max} / 10$ or $|d_n| > 10d_{max}$ respectively, where d_{max} is the preset maximum desired value for d_n stored in cell D2. The step-size change factor 2 is used for the following reason: for the improved polygonal Euler method, d_n is $O(h^3)$, and so an increase in d_n by a factor of 10 will correspond to an increase in h by a factor $\sqrt[3]{10} \approx 2.1 \approx 2$.

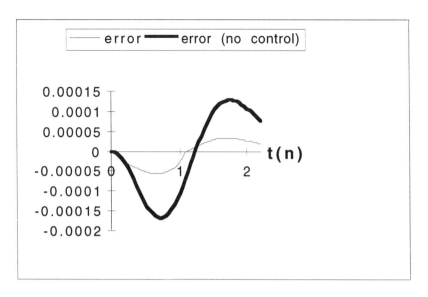

Graph 6.11.2 *The error with and without step-size control when initially h = 0.04.*

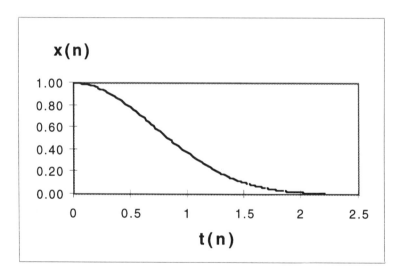

Graph 6.11.3 *The solution, h = 0.04.*

Exercises 6.11

(1) Experiment with the spreadsheet of Example 6.11.1, varying the initial step-size and the value of d_{max}.

(2) Apply the method used in Example 6.11.1 to the problem

$$\frac{dx}{dt} = 2\sqrt{x-1}, \quad x(0) = \alpha > 1, \text{ exact solution } x = (t + \sqrt{\alpha-1})^2 + 1$$

of Example 6.1.4, and compare the results with those obtained there with the modified Euler method.

(3) Van der Pol's equation (see Exercise 6.7(2)) describing a relaxation oscillator is equivalent to the 1st-order pair

$$\dot{x} = y$$
$$\dot{y} = -ax + \varepsilon(1 - x^2)y$$

Extend the improved polygonal Euler method and Kutta's 3rd-order method to a system of two equations, so that, for example, the magnitude of the estimated local truncation error is given by

$$\|\mathbf{d}_{k+1}\| \approx \tfrac{1}{6}h\sqrt{(k_1 - 2k_2 + k_3)^2 + (l_1 - 2l_2 + l_3)^2} \ .$$

Apply this method of step-size control to the pair above, beginning with the parameter and initial values $a = 2$, $\varepsilon = 2.5$, $x(0) = 1.5$, $y(0) = 1$. Use an initial step-size $h = 0.05$, with $\|\mathbf{d}\|_{max} = 0.0001$ and compute 250 steps. The values obtained could be checked by also calculating them with the 4th-order Runge-Kutta method with step-size 0.1, as in Exercise 6.7(2).

(4) Repeat the preceding Exercise for the Lotka-Volterra equations described in Exercise 6.6(3).

7 Partial Differential Equations

Many of the differential equations that occur in science and engineeering involve quantities that are dependent on two or more independent variables. We will consider the numerical solution of some second-order partial differential equations (PDE's) with two independent variables which have the general form

$$a\frac{\partial^2 u}{\partial x^2} + b\frac{\partial^2 u}{\partial x \partial y} + c\frac{\partial^2 u}{\partial y^2} + d\frac{\partial u}{\partial x} + e\frac{\partial u}{\partial y} + fu + g = 0$$

Here, u is a function of x and y and the coefficients $a, b, c, d, e, f,$ and g may be functions of x, y and u.

PDE's of this type are classified - according to the values of the quantity $\Delta = b^2 - 4ac$ - as being *elliptic* if $\Delta < 0$, *parabolic* if $\Delta = 0$, or *hyperbolic* if $\Delta > 0$. Note that the class of a particular PDE may depend on the region over which the PDE is being solved. We will not concern ourselves any further with the consequences of this classification, but we will look at important simple examples of each type.

§7.1 Elliptic equations

Poisson's equation is an important example of the elliptic type. It arises in many contexts including electromagnetism (relating potential to charge distribution), gravitation, torsion theory, and fluid mechanics. In two dimensions it has the form

$$\frac{\partial^2 u(x,y)}{\partial x^2} + \frac{\partial^2 u(x,y)}{\partial y^2} = g(x,y)$$

where (x,y) lies in a region S that is bounded by a closed curve C.

If the right-hand side is zero, $g(x,y) = 0$, then the PDE is called **Laplace's equation**. This equation describes the equilibrium distribution of a diffusible quantity (e.g. temperature) over a two-dimensional domain.

Boundary conditions must also be specified, with either $u(x,y)$ or the normal derivative of u being given at every point on C. For example, S might be the interior of the rectangular region $\{(x,y)|\, a \le x \le b, c \le y \le d\}$, with the boundary conditions to be provided being the values of $u(a,c), u(a,d), u(b,c)$, and $u(b,d)$.

As with ordinary differential equations, we need to partition the domain S into a discrete mesh of points. For a rectangular domain, one way is to subdivide $[a,b]$ into M sub-intervals of size $h = (b-a)/M$, and $[c,d]$ into N sub-intervals of size $k = (d-c)/N$. The mesh points are then the points (x_i, y_j), where

$$x_i = a + ih, \quad i = 0, 1, 2, ..., M, \, and \, y_j = c + jk, \quad j = 0, 1, 2, ..., N.$$

The function $u(x,y)$ is represented on this mesh by the discrete values $u_{i,j} = u(x_i, y_j)$. Working as in §6.8, Taylor's theorem can be used to show that

$$\frac{\partial^2 u}{\partial x^2}(x_i, y_j) = \frac{u_{i+1,j} - 2u_{i,j} + u_{i-1,j}}{h^2} - \frac{h^2}{12} \frac{\partial^4 u}{\partial x^4}(\xi_i, y_j)$$

and

$$\frac{\partial^2 u}{\partial y^2}(x_i, y_j) = \frac{u_{i,j+1} - 2u_{i,j} + u_{i,j-1}}{k^2} - \frac{h^2}{12}\frac{\partial^4 u}{\partial y^4}(x_i, \eta_j)$$

where

$$x_{i-1} < \xi_i < x_{i+1} \quad and \quad y_{j-1} < \eta_j < y_{j+1}.$$

Thus the use of central difference approximations for the derivatives leads to the following difference equation approximation for the Poisson equation:

$$\frac{w_{i+1,j} - 2w_{i,j} + w_{i-1,j}}{h^2} + \frac{w_{i,j+1} - 2w_{i,j} + w_{i,j-1}}{k^2} = g(x_i, y_j) \qquad (7.1.1)$$

Here, the values of $u_{i,j}$ are approximated by $w_{i,j}$ at the interior mesh points and are determined by the difference scheme (7.1.1) and the boundary conditions.

Note: The scheme (7.1.1) relates $w_{i,j}$ to the values at the 4 nearest sites of the mesh. If we write the value of $w_{i,j}$ at the centre as w_c and the values at the 4 points (**t**op, **l**eft, **b**ottom & **r**ight) of this "star" as w_t, w_l, w_b and w_r, as shown in the diagram below,

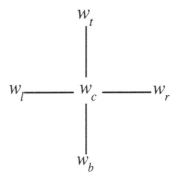

then (7.1.1) becomes

$$\frac{w_r - 2w_c + w_l}{h^2} + \frac{w_t - 2w_c + w_b}{k^2} = g(x_c, y_c)$$

and with a square mesh, $h = k$, we get the simpler form

$$\frac{w_t + w_l + w_b + w_r - 4w_c}{h^2} = g(x_c, y_c) \qquad (7.1.2)$$

The system of linear equations (7.1.1) can be rendered into a more convenient form by relabelling the mesh points so as to create an equivalent 1-dimensional mesh. In particular cases the presence of any **symmetries** in the geometry of S and in the boundary conditions may be used to reduce the number of unknowns.

It is only really convenient to perform numerical solutions on a spreadsheet for those methods which result either in an explicit relationship (rather than an implicit one), or in a set of linear equations that can be solved by means of an iterative method. These points will be demonstrated in the examples that follow.

7.1.1 Example

Solve Laplace's equation

$$\frac{\partial^2 u(x,y)}{\partial x^2} + \frac{\partial^2 u(x,y)}{\partial y^2} = 0$$

on the domain $\{(x,y) \mid 0 < x < 1, 0 < y < 1\}$ *with the boundary conditions*

$$u(x,0) = 0 = u(0,y), \quad u(x,1) = x, \quad u(1,y) = y.$$

The exact solution is $u(x,y) = xy$.

Note that the boundary conditions are symmetric about the line $y = x$, as is (trivially in this case) the right-hand side of the PDE. In order to exploit this symmetry and minimise the number of equations to be solved we should choose $M = N$ i.e. $k = h$, so that $w_{i,j} = w_{j,i}$.

If we take $M = 5$ this will result in $\frac{1}{2}M(M-1) = 10$ equations, instead of $(M-1)^2 = 16$ (in the case of no symmetry). Using the symmetry roughly halves the number of equations to be solved, an economy that is more easily appreciated when the mesh is finer, e.g. when $M = 100$.

We can label the interior sites P_i as shown in Fig.7.1.1 below.

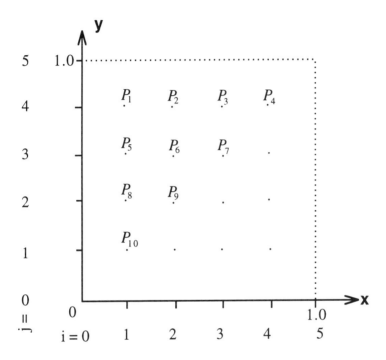

Fig.7.1.1 *Symmetry allows a reduction of the number of mesh points, in this case from 16 to 10 points. A suitable relabelling of these points is shown here.*

Now, for convenience, we will let the $w_{i,j}$ at the interior sites be written as w_k at site P_k. For example at P_8 where $i = 1$ and $j = 2$ we write w_8 instead of $w_{1,2}$.

The boundary conditions become $w_{0,m} = w_{m,0} = 0, m = 0, 1, ..., 5$ and $w_{m,5} = x_m = mh, w_{5,m} = y_m = mh, m = 1, 2, ..., 5$.

With $M = 5$ we have $h = 0.2$, and if (7.1.2) is used at site P_1 the result is

$$w_{1,5} + w_{4,0} + w_5 + w_2 - 4w_1 = h^2 g(x_1, y_4) = 0$$

$$\Rightarrow -4w_1 + w_2 + w_5 = -(w_{1,5} + w_{4,0}) = -x_1 = -h = -0.2.$$

At site P_2 we get $w_1 - 4w_2 + w_3 + w_6 = -w_{2,5} = -x_2 = -2h = -0.4$, and so on. The equations at the rest of the sites are shown below.

Site	Equation
P_3	$w_2 - 4w_3 + w_4 + w_7 = -w_{3,5} = -x_3 = -3h = -0.6$
P_4	$w_3 - 2w_4 = \frac{1}{2}(-w_{4,5} - w_{5,4}) = \frac{1}{2}(-x_3 - y_4) = -4h = -0.8$
P_5	$w_1 - 4w_5 + w_6 + w_8 = -w_{0,3} = 0$
P_6	$w_2 + w_5 - 4w_6 + w_7 + w_9 = 0$
P_7	$w_3 + w_6 - 2w_7 = 0$
P_8	$w_5 - 4w_8 + w_9 + w_{10} = -w_{0,2} = 0$
P_9	$w_6 + w_8 - 2w_9 = 0$
P_{10}	$w_8 - 2w_{10} = \frac{1}{2}(-w_{0,1} - w_{1,0}) = 0$

We now have a system of 10 linear equations to be solved. On a spreadsheet this is best done using an iterative method such as the Gauss-Seidel method. From the equations listed above we get

$$w_1^{(n+1)} = \tfrac{1}{4}[0.2 + w_2^{(n)} + w_5^{(n)}]$$

$$w_2^{(n+1)} = \tfrac{1}{4}[0.4 + w_1^{(n+1)} + w_3^{(n)} + w_6^{(n)}]$$

$$w_3^{(n+1)} = \tfrac{1}{4}[0.6 + w_2^{(n+1)} + w_4^{(n)} + w_7^{(n)}]$$

$$w_4^{(n+1)} = \tfrac{1}{2}[0.8 + w_3^{(n+1)}]$$

$$w_5^{(n+1)} = \tfrac{1}{4}[w_1^{(n+1)} + w_6^{(n)} + w_8^{(n)}]$$

$$w_6^{(n+1)} = \tfrac{1}{4}[w_2^{(n+1)} + w_5^{(n+1)} + w_7^{(n)} + w_9^{(n)}]$$

$$w_7^{(n+1)} = \tfrac{1}{2}[w_3^{(n+1)} + w_6^{(n+1)}]$$

$$w_8^{(n+1)} = \tfrac{1}{4}[w_5^{(n+1)} + w_9^{(n)} + w_{10}^{(n)}]$$

$$w_9^{(n+1)} = \tfrac{1}{2}[w_6^{(n+1)} + w_8^{(n+1)}]$$

$$w_{10}^{(n+1)} = \tfrac{1}{2}w_8^{(n+1)}$$

A suitable spreadsheet is shown in Screen 7.1.1. The 10 iterations are tabulated in columns B to K, with the starting values $w_i = 0.5$, $i = 1, 2, ..., 10$ in row 7. The corresponding exact values are calculated in row 3.

Note that after about 48 steps the exact solution is achieved by the iteration. This is to be expected in this example since

$$\frac{\partial^4 u}{\partial x^4} = \frac{\partial^4 u}{\partial y^4} = 0$$

and therefore the truncation error is zero.

	A	B	C	D
1	Example		Laplaces equation	
2	7.1.1		Elliptic PDE	
3	Exact=	=0.16	=0.32	=0.48
4				
5	Gauss-	Seidel	iteration	
6	n	w1(n)	w2(n)	w3(n)
7	=0	=0.5	=0.5	=0.5
8	=A7+1	=(0.2+C7+F7)/4	=(0.4+B8+D7+G7)/4	=(0.6+C8+E7+H7)/4

	E	F	G
3	=0.64	=0.12	=0.24
4			
5			
6	w4(n)	w5(n)	w6(n)
7	=0.5	=0.5	=0.5
8	=(0.8+D8)/2	=(B8+G7+I7)/4	=(C8+F8+H7+J7)/4

	H	I	J	K
3	=0.36	=0.08	=0.16	=0.04
4				
5				
6	w7(n)	w8(n)	w9(n)	w10(n)
7	=0.5	=0.5	=0.5	=0.5
8	=(D8+G8)/2	=(F8+J7+K7)/4	=(G8+I8)/2	=I8/2

Screen 7.1.1 *Solving Laplace's equation. The symmetry of the domain and boundary conditions about y = x has been used to reduce the number of equations to be solved by Gauss-Seidel iteration.*

	A	B	C	D
1	Example		Laplaces equation	
2	7.1.1		Elliptic PDE	
3	Exact=	0.16	0.32	0.48
4				
5	Gauss-	Seidel	iteration	
6	n	ψ1(n)	ψ2(n)	ψ3(n)
7	0	0.5	0.5	0.5
8	1	0.3	0.425	0.50625
9	2	0.2375	0.3953125	0.530078125

	E	F	G	H
3	0.64	0.12	0.24	0.36
4				
5				
6	ψ4(n)	ψ5(n)	ψ6(n)	ψ7(n)
7	0.5	0.5	0.5	0.5
8	0.653125	0.325	0.4375	0.471875
9	0.665039063	0.2515625	0.37578125	0.452929688

	I	J	K
3	0.08	0.16	0.04
4			
5			
6	ψ8(n)	ψ9(n)	ψ10(n)
7	0.5	0.5	0.5
8	0.33125	0.384375	0.165625
9	0.200390625	0.288085938	0.100195313

Screen 7.1.1 (continued)

7.1.2 Derivative Boundary Conditions

An elliptic equation may occur, for example, as the steady-state version of a (parabolic) heat-diffusion equation. In such cases the temperature *gradient* may be specified at part of the boundary rather than the temperature itself. For example the temperature gradient may be proportional to the difference between the temperature at the boundary and the "external" temperature, with the constant of proportionality representing the degree of insulation at the boundary.

7.1.3 Example

Consider Laplace's equation

$$\frac{\partial^2 u(x,y)}{\partial x^2} + \frac{\partial^2 u(x,y)}{\partial y^2} = 0$$

on the domain $\{(x,y)|\, 0<x<1, 0<y<1\}$ *with the boundary conditions* $u(x,0) = 1 = u(0,y)$, *plus*

$$\frac{\partial u}{\partial x}(1,y) = -\alpha u(1,y),\ 0 \le y \le 1, \text{ and } \frac{\partial u}{\partial y}(x,1) = -\alpha u(x,1),\ 0 \le x \le 1.$$

Again we have symmetry about $y = x$, and we take $M = N = 5$, i.e. $h = k = 0.2$. The mesh will now be extended to the fictitious sites P_6', P_7', P_8', P_9', and P_4', as shown in Fig.7.1.2. These sites facilitate the numerical approximation of the derivative boundary condition.

The PDE now applies along the edges $x = 1, 0 < y \le 1$ and $y = 1, 0 < x \le 1$ as well as in the interior of the domain, and the central difference approximations for the derivative boundary conditions will allow the elimination of all terms involving the fictitious sites from the difference approximation for the PDE.

From $u(x,0) = 1 = u(0,y)$ we have $w_{0,m} = w_{m,0} = 1, m = 0, 1, ..., 5$, the boundary values specified along the left and bottom edges of the domain.

When (7.1.2) is applied at P_1 we get $w_6' + 1 + w_6 + w_2 - 4w_1 = 0$. At the same point the boundary condition is

$$\frac{\partial u}{\partial y}(x,1) = -\alpha u(x,1) \Rightarrow \frac{w_6' - w_6}{2k} \approx -\alpha w_1$$

when a central difference approximation is used for the derivative.

Hence $w_6' - w_6 = -2\alpha k w_1$. After w_6' is eliminated we get

$$(4 + 2\alpha k)w_1 - w_2 - 2w_6 = 1.$$

A similar process can be applied at P_2, P_3, P_4, and P_5, giving the equations

$$(4 + 2\alpha k)w_2 - w_1 - w_3 - 2w_7 = 0$$
$$(4 + 2\alpha k)w_3 - w_2 - w_4 - 2w_8 = 0$$
$$(4 + 2\alpha k)w_4 - w_3 - w_5 - 2w_9 = 0$$
$$(1 + \alpha k)w_5 - w_4 = 0$$

and the direct use of (7.1.2) alone is all that is needed at sites P_6 to P_{15}, as before.

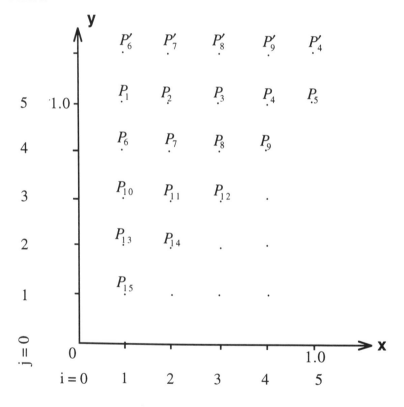

Fig.7.1.2 *The top edge of the mesh is extended here to include fictitious sites in order to cater for derivative boundary conditions.*

The final set of 15 linear equations can then be solved using the Gauss-Seidel iteration. Screen 7.1.2 shows parts of a spreadsheet designed to do this, with the results (with $\alpha = 1$) at the 100th iteration duplicated in the meshed form in the block Q2 to W8, for convenience. This block can be used to generate a graphical display, for example the 3-dimensional "wireframe" shown in Graph 7.1.1.

	A	B	C	D
1	Example	Elliptic PDE	Laplace's	equation
2	7.1.3	Derivative	alpha =	=0.5
3		b/c's	A =	=1/(4+0.4*D2)
4			B =	=1/(1+0.2*D2)
5	Gauss-	Seidel	iteration	
6	n	v1(n)	v2(n)	v3(n)
7	=0	=1	=1	=1
8	=A7+1	=D$3*(1+C7+2*G7)	=D$3*(B8+D7+2*H7)	=D$3*(C8+E7+2*I7)

	E	F	G
6	v4(n)	v5(n)	v6(n)
7	=1	=1	=1
8	=D$3*(D8+F7+2*J7)	=D$4*E8	=(1+B8+H7+K7)/4

	H	I	J
6	v7(n)	v8(n)	v9(n)
7	=1	=1	=1
8	=(C8+G8+I7+L7)/4	=(D8+H8+J7+M7)/4	=(E8+I8)/2

	K	L	M
6	v10(n)	v11(n)	v12(n)
7	=1	=1	=1
8	=(1+G8+L7+N7)/4	=(H8+K8+M7+O7)/4	=(I8+L8)/2

	N	O	P
6	v13(n)	v14(n)	v15(n)
7	=1	=1	=1
8	=(1+K8+O7+P7)/4	=(L8+N8)/2	=(1+N8)/2

	Q	R	S	T	U	V	W
2		x=0	x=0.2	x=0.4	x=0.6	x=0.8	x=1
3	y=0	=1	=1	=1	=1	=1	=1
4	y=0.2	=1	=P107	=N107	=K107	=G107	=B107
5	y=0.4	=1	=N107	=O107	=L107	=H107	=C107
6	y=0.6	=1	=K107	=L107	=M107	=I107	=D107
7	y=0.8	=1	=G107	=H107	=I107	=J107	=E107
8	y=1	=1	=B107	=C107	=D107	=E107	=F107

Screen 7.1.2 *Solving Laplace's equation with derivative boundary conditions.*

	A	B	C	D	E	F	G	H
1	Example	Elliptic PDE	Laplace's	equation				
2	7.1.3	Derivative	alpha =	0.5				
3		b/c's	A =	0.2381				
4			B =	0.9091				
5	Gauss-	Seidel	iteration					
6	n	1(n)	2(n)	3(n)	4(n)	5(n)	6(n)	7(n)
7	0	1	1	1	1	1	1	1
8	1	0.952	0.941	0.938	0.938	0.852	0.988	0.982

	I	J	K	L	M	N	O	P
6	8(n)	9(n)	10(n)	11(n)	12(n)	13(n)	14(n)	15(n)
7	1	1	1	1	1	1	1	1
8	0.980	0.959	0.997	0.995	0.987	0.999	0.997	1.000

Screen 7.1.2 (continued)

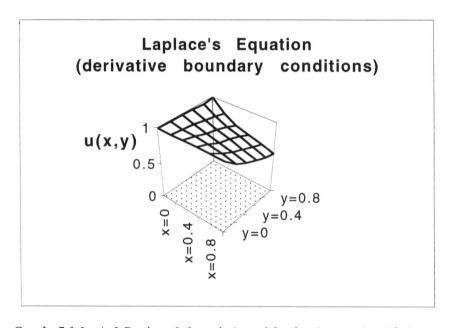

Graph 7.1.1 A 3-D plot of the solution of Laplace's equation (derivative boundary conditions).

Exercises 7.1

(1) Derive the difference approximations (at sites $P_2, P_3, ..., P_{10}$) quoted above in Example 7.1.1.

(2) Find a numerical approximation for the solution of the elliptic PDE

$$\frac{\partial^2 u}{\partial x^2} + \frac{\partial^2 u}{\partial y^2} = xe^y \text{ on the region } \{(x,y)| 0 < x < 1, 0 < y < 1\},$$

subject to the boundary conditions

$$u(0,y) = 0, \quad u(1,y) = e^y, \quad 0 \le y \le 1$$
$$u(x,0) = x, \quad u(x,1) = ex, \quad 0 \le x \le 1$$

The exact solution is $u(x,y) = xe^y$. Use $M = 4, N = 5$.

(3) Derive all the difference equations needed in Example 7.1.3, set up the spreadsheet, and investigate the effect of varying the "insulation" parameter α.

(4) Repeat the problem of Example 7.1.3 but with the possibility of different "insulation" along the edge $y = 1$, so that

$$\frac{\partial u}{\partial y}(x,1) = -\beta u(x,1), 0 \le x \le 1 \text{ with } \alpha \ne \beta.$$

The problem is no longer symmetric about $y = x$ so the mesh points previously omitted due to symmetry must now be included, plus some more fictitious points adjacent to the right-hand edge.

§7.2 Parabolic Equations

The classic example is the heat (or diffusion) equation

$$\kappa \frac{\partial u}{\partial t} - \frac{\partial^2 u}{\partial x^2} = 0, \text{ with } \kappa > 0 \text{ on } 0 < x < L, \quad t > 0,$$

with the boundary conditions $u(0,t) = \phi(t)$, $u(L,t) = \psi(t)$, and the initial condition $u(x,0) = f(x), 0 \le x \le L$, with $f(0) = \phi(0), f(L) = \psi(0)$.

Taking $t_n = nk, n = 0, 1, 2, \ldots$, Taylor's theorem gives

$$\frac{\partial u}{\partial t}(x_i, t_j) = \frac{u(x_i, t_j + k) - u(x_i, t_j)}{k} - \frac{k}{2} \frac{\partial^2 u}{\partial t^2}(x_i, \mu_j)$$

$$\approx \frac{w_{i,j+1} - w_{i,j}}{k}$$

where $t_j < \mu_j < t_{j+1}$. When this forward difference approximation is combined with the Taylor theorem result for the second x-derivative (found in §7.1), we get a difference approximation for the heat equation at the interior sites (referring to the variable x) of the mesh:

$$\kappa \frac{w_{i,j+1} - w_{i,j}}{k} - \frac{w_{i+1,j} - 2w_{i,j} + w_{i-1,j}}{h^2} = 0,$$

$$\Rightarrow w_{i,j+1} = \left(1 - \frac{2k}{\kappa h^2}\right) w_{i,j} + \frac{k}{\kappa h^2}(w_{i+1,j} + w_{i-1,j}) \qquad (*)$$

for $i = 1, 2, \ldots, M-1, \quad j = 0, 1, 2, \ldots$.

On the mesh the boundary conditions are

$$w_{0,j} = \phi(t_j), \quad w_{M,j} = \psi(t_j), \quad j = 0, 1, 2, \ldots$$

and the initial condition becomes $w_{i,0} = f(x_i), \quad i = 0, 1, 2, \ldots, M$.

If we define $\lambda = \dfrac{k}{\kappa h^2}$, then the difference equations $(*)$ are

$$w_{i,j+1} = \lambda w_{i-1,j} + (1 - 2\lambda)w_{i,j} + \lambda w_{i+1,j} \qquad (7.2.1)$$

These equations ($M-1$ of them) can be put in matrix form:

$$\mathbf{w}^{(j+1)} = \mathbf{A}\mathbf{w}^{(j)} + \mathbf{b}_j$$

where $\mathbf{A}$ is an $(M-1) \times (M-1)$ matrix, and

$$\mathbf{w}^{(j)} = \begin{bmatrix} w_{1,j} \\ w_{2,j} \\ w_{3,j} \\ \vdots \\ w_{M-1,j} \end{bmatrix}, \quad \mathbf{A} = \begin{bmatrix} 1-2\lambda & \lambda & 0 & 0 & \cdots & 0 \\ \lambda & 1-2\lambda & \lambda & 0 & \cdots & 0 \\ 0 & \lambda & 1-2\lambda & \lambda & \cdots & 0 \\ \vdots & \vdots & \vdots & \vdots & \vdots & \vdots \\ 0 & 0 & & \cdots & 0 & \lambda & 1-2\lambda \end{bmatrix}$$

and

$$\mathbf{b}_j = \begin{bmatrix} \lambda w_{0,j} \\ 0 \\ \vdots \\ 0 \\ \lambda w_{M,j} \end{bmatrix}.$$

Thus the vector $\mathbf{w}^{(j+1)}$ at t_{j+1} is obtained explicitly from $\mathbf{w}^{(j)}$.

It would seem that the truncation error should be $O(k+h^2)$. This is the case, but the stability of the iteration $\mathbf{w}^{(j+1)} = \mathbf{A}\mathbf{w}^{(j)} + \mathbf{b}_j$ requires the extra condition $\lambda \leq \frac{1}{2}$. Thus the convergence (as $k \to 0$) to the exact solution is *conditionally stable*.

7.2.1 Example

Solve

$$\frac{\partial u}{\partial t} - \frac{\partial^2 u}{\partial x^2} = 0 \text{ , on } 0 < x < 1, \quad t > 0, \text{with } u(0,t) = u(1,t) = 0, t > 0$$

and $u(x,0) = \sin(\pi x), 0 \le x \le 1$. *The exact solution is*

$$u(x,t) = \exp(-\pi^2 t)\sin(\pi x).$$

Here, $\kappa = 1$, hence the stability condition is $k \le \frac{1}{2}h^2$. With $M = 5$ i.e. $h = 0.2$ this is $k \le 0.02$. The boundary conditions indicate that $\mathbf{b}_j = \mathbf{0}$, and the initial condition becomes

$$w_{i,0} = \sin(\pi x_i) = \sin(\pi i h), i = 0, 1, 2, ..., M.$$

With this rather coarse mesh the matrix $\mathbf{A}$ is 4×4. In fact the symmetry of the boundary and initial conditions about $x = 0.5$ means we need only solve on $0 \le x \le 0.5$, an economy that doesn't matter with the small value of M used here but which could be significant when M is very large.

Screen 7.2.1 shows a suitable spreadsheet for this problem. The quantity $1 - 2\lambda$ is stored in cell D4 in order to simplify the formulas in columns D, E, F & G. In order to show (a) the effects of different values of λ and (b) what happens if the stability condition is not satisfied, the spreadsheet includes columns tabulating the exact solution $u(0.4,t)$ at $x = 0.4$ and the percentage error in the numerical estimate $w_{2,j}$ of this quantity (see the first exercise below).

The method discussed above used the **forward** difference approximation for the first t-derivative,

$$\frac{\partial u}{\partial t}(x_i, t_j) \approx \frac{w_{i,j+1} - w_{i,j}}{k},$$

and the stability of the iteration to large t was conditional, requiring that

$$\frac{k}{\kappa h^2} \le \frac{1}{2}.$$

	A	B	C	D
1	Example	7.2.1	h =	=0.2
2	Parabolic		k =	=0.01
3	PDE		λ = k/h^2 =	=D2/D1^2
4			1−2λ =	=1−2*D3
5		x =	=0	=C5+$D1
6	j	t(j)	⩗(0,j)	⩗(1,j)
7	=0	=0	=SIN(PI()*C5)	=SIN(PI()*D5)
8	=A7+1	=B7+D$2	=0	=$D$4*D7+$D$3*E7

	E	F
5	=D5+$D1	=E5+$D1
6	⩗(2,j)	⩗(3,j)
7	=SIN(PI()*E5)	=SIN(PI()*F5)
8	=$D$3*D7+$D$4*E7+$D$3*F7	=$D$3*E7+$D$4*F7+$D$3*G7

	G	H
5	=F5+$D1	=G5+$D1
6	⩗(4,j)	⩗(5,j)
7	=SIN(PI()*G5)	=SIN(PI()*H5)
8	=$D$3*F7+$D$4*G7	=0

	I	J
5	exact	%
6	u(0.4,t)	error
7	=EXP(−PI()^2*B7)*SIN(PI()*0.4)	=100*(I7−E7)/I7

	A	B	C	D
1	Example	7.2.1	h =	0.2
2	Parabolic		k =	0.01
3	PDE		λ = k/h^2 =	0.25
4			1−2λ =	0.5
5		x =	0	0.2
6	j	t(j)	⩗(0,j)	⩗(1,j)
7	0	0	0	0.588
8	1	0.01	0	0.532

	E	F	G	H	I	J
5	0.4	0.6	0.8	1	exact	%
6	⩗(2,j)	⩗(3,j)	⩗(4,j)	⩗(5,j)	u(0.4,t)	error
7	0.951	0.951	0.588	1E−16	0.9511	0.0
8	0.860	0.860	0.532	0	0.8617	0.2

Screen 7.2.1 *Solution of the heat equation using a forward diference approximation for the time derivative, requiring an upper bound on the time step-size, for stability.*

267

If instead the **backward** difference approximation

$$\frac{\partial u}{\partial t}(x_i, t_j) \approx \frac{w_{i,j} - w_{i,j-1}}{k}$$

is used then the resulting difference approximation is unconditionally stable. The difference formula is now

$$-\lambda w_{i-1,j} + (1 + 2\lambda) w_{i,j} - \lambda w_{1+1,j} = w_{i,j-1}$$

at the interior sites, $i = 1, 2, ..., M-1$, $j = 1, 2, ...$ This system of equations has the matrix form

$$\mathbf{B}\mathbf{w}^{(j)} = \mathbf{w}^{(j-1)} + \mathbf{b}_j$$

where the vector $\mathbf{b}_j$ is as defined earlier and $\mathbf{B}$ is the tri-diagonal matrix

$$\mathbf{B} = \begin{bmatrix} 1+2\lambda & -\lambda & 0 & 0 & \cdots & 0 \\ -\lambda & 1+2\lambda & -\lambda & 0 & \cdots & 0 \\ 0 & -\lambda & 1+2\lambda & -\lambda & \cdots & \\ \vdots & & & & & \\ 0 & 0 & 0 & \cdots & -\lambda & 1+2\lambda \end{bmatrix}.$$

For each t-step, this system can be solved either using the method for tri-diagonal systems discussed in Chapter 3 (Crout's algorithm), or (since the matrix is strictly diagonally dominant) with the Gauss-Seidel or SOR iterative methods. Unfortunately neither is conveniently implemented on a spreadsheet in the present context.

Further improvements are easily achieved, for example to make the truncation error $O(k^2 + h^2)$ with *unconditional* stability (the Crank-Nicholson method). Again the difference equations are implicit and not conveniently handled on a spreadsheet.

Exercises 7.2

(1) Explore the effect on the stability and error in Example 7.2.1 by computing the value of the numerical approximation for $u(0.4, 0.4)$ using the following values of k: 0.06, 0.04, 0.02, 0.01, 0.005, 0.0025.

(2) Modify the spreadsheet so that $M = 10$, repeat the calculations of the previous problem, and compare the results with those found with $M = 5$.

(3) Solve

$$\pi^2 \frac{\partial u}{\partial t} - \frac{\partial^2 u}{\partial x^2} = 0 \text{ on } 0 < x < 2, \quad t > 0$$

with $u(0, t) = u(2, t) = 0$, $t > 0$, and

$$u(x, 0) = \sin(\tfrac{1}{2}\pi x)(1 + 2\cos(\tfrac{1}{2}\pi x)), \quad 0 \le x \le 1.$$

The exact solution is $u(x, t) = e^{-t}\sin(\pi x) + e^{-t/4}\sin(\tfrac{1}{2}\pi x)$. With $h = 0.2$ compare the numerical solution with the exact solution for $u(1, 0.5)$ (for example) using the following values of k: 0.2, 0.1, 0.05, 0.02.

§7.3 Hyperbolic Equations

A common and important example is the 1-dimensional wave equation

$$\frac{\partial^2 u}{\partial t^2} - c^2 \frac{\partial^2 u}{\partial x^2} = 0 \text{ on } 0 < x < L, \quad t > 0,$$

where c is a constant (the wave velocity), with boundary conditions

$$u(0,t) = u(L,t) = 0, \quad t > 0$$

and the initial conditions

$$u(x,0) = f(x) \text{ and } u_t(x,0) = \frac{\partial u}{\partial t}(x,0) = g(x), \quad 0 \le x \le L.$$

Using central difference approximations for the second derivatives (as in §7.1) we get the difference equation

$$\frac{w_{i,j+1} - 2w_{i,j} + w_{i,j-1}}{k^2} - c^2 \frac{w_{i+1,j} - 2w_{i,j} + w_{i-1,j}}{h^2} = 0$$

for $w_{i,j}$ at the interior points of the mesh,

$$i = 1, 2, ..., M-1, \quad j = 1, 2, ...,$$

with truncation error $O(k^2 + h^2)$.

Let $\mu = ck / h$ and solve for $w_{i,j+1}$:

$$w_{i,j+1} = \mu^2 w_{i+1,j} + 2(1 - \mu^2)w_{i,j} + \mu^2 w_{i-1,j} - w_{i,j-1} \tag{7.3.1}.$$

The boundary conditions give $w_{0,j} = w_{M,j} = 0, j = 1, 2, ...$ and the initial condition $u(x,0) = f(x)$ gives $w_{i,0} = f(x_i), i = 0, 1, 2, ..., M$.

The equations (7.3.1) can be put in matrix form:

$$\mathbf{w}^{(j+1)} = \mathbf{Cw}^{(j)} - \mathbf{w}^{(j-1)} \tag{7.3.2}$$

where $\mathbf{C}$ is the $(M-1) \times (M-1)$ matrix

$$C = \begin{bmatrix} 2(1-\mu^2) & \mu^2 & 0 & 0 & \cdots & 0 \\ \mu^2 & 2(1-\mu^2) & \mu^2 & 0 & \cdots & 0 \\ 0 & \mu^2 & 2(1-\mu^2) & \mu^2 & \cdots & 0 \\ \vdots & & & & & \\ 0 & 0 & 0 & \cdots & \mu^2 & 2(1-\mu^2) \end{bmatrix}$$

The explicit formula (7.3.2) requires *two* starting values, $\mathbf{w}^{(0)}$ and $\mathbf{w}^{(1)}$. The former is given by $w_{i,0} = f(x_i)$ and the latter must be obtained from the other initial condition, $u_t(x,0) = g(x)$. The simplest (and crudest) way is to use the forward difference approximation for the time derivative, leading to

$$w_{i,1} = w_{i,0} + kg(x_i), \quad i = 1, 2, \ldots, M-1 \tag{7.3.3}$$

with the disadvantage that the overall truncation error is now $O(k + h^2)$. This can be restored to $O(k^2 + h^2)$ if the wave equation is assumed to be valid along the initial line, in which case Taylor's Theorem gives

$$w_{i,1} = w_{i,0} + kg(x_i) + \frac{c^2 k^2}{2} f''(x_i) \tag{7.3.4}$$

In the case where $f''(x_i)$ is not available it can be replaced with a central difference approximation (assuming f is at least 4 times differentiable on $[0,L]$), so that

$$w_{i,1} = (1-\mu^2) f(x_i) + \tfrac{1}{2}\mu^2 [f(x_{i+1}) + f(x_{i-1})] + kg(x_i) \tag{7.3.5}$$

The explicit algorithm (7.3.2) is conditionally stable with truncation error $O(k^2 + h^2)$, provided that $\mu \le 1$ and that f and g are sufficiently differentiable.

Implicit and unconditionally stable methods exist for hyperbolic equations, but they are not conveniently implemented on a spreadsheet.

7.3.1 Example

Solve

$$\frac{\partial^2 u}{\partial t^2} - \frac{\partial^2 u}{\partial x^2} = 0 \ \ on \ \ 0 < x < 1, \ \ t > 0$$

with $u(0,t) = u(1,t) = 0, \ \ t > 0 \ \ and \ \ u(x,0) = \sin(\pi x), \ and$

$$\frac{\partial u}{\partial t}(x,0) = \pi \sin(\pi x) \ \ for \ \ 0 \le x \le 1.$$

This has exact solution $u(x,t) = \sin(\pi x)[\sin(\pi t) + \cos(\pi t)].$

In this example $c = 1$. If we use $M = 5$ i.e. $h = 0.2$ then the stability condition is $k \le 0.2$. Since $f(x) = \sin(\pi x)$ and $g(x) = \pi \sin(\pi x)$ then in this case $w_{i,0} = \sin(\pi x_i)$ and (7.3.4) gives

$$w_{i,1} = w_{i,0} + k\pi \sin(\pi x_i) - \frac{k^2 \pi^2}{2} \sin(\pi x_i)$$

$$= [1 + k\pi - \tfrac{1}{2}(k\pi)^2] \sin(\pi x_i)$$

Screen 7.3.1 shows a spreadsheet for this example, and Graph 7.3.1 plots the exact and numerical solutions for $u(0.4,t)$ on $0 \le t \le 5$, using $k = 0.1$. For convenience the quantities μ^2 and $2(1 - \mu^2)$ are stored in cells D4 and D5.

	A	B	C	D
1	**Example**		**h** =	=0.2
2	**7.3.1**		**k** =	=0.1
3	**Hyperbolic**		**μ** =	=D2/D1
4	**PDE**		**μ^2** =	=D3^2
5	k*π =	=D2*PI()	**2(1-μ^2)** =	=2*(1-D4)
6		x =	=0	=C6+$D1
7	**j**	**t(j)**	**w(0,j)**	**w(1,j)**
8	=0	=0	=0	=SIN(PI()*D6)
9	=A8+1	=B8+D$2	=0	=D8*(1+$B5-0.5*$B5^2)
10	=A9+1	=B9+D$2	=0	=$D$5*D9+$D$4*E9-D8

	E	F
6	=D6+$D1	=E6+$D1
7	**w(2,j)**	**w(3,j)**
8	=SIN(PI()*E6)	=SIN(PI()*F6)
9	=E8*(1+$B5-0.5*$B5^2)	=F8*(1+$B5-0.5*$B5^2)
10	=$D$4*D9+$D$5*E9+$D$4*F9-E8	=$D$4*E9+$D$5*F9+$D$4*G9-F8

	G	H
6	=F6+$D1	=G6+$D1
7	**w(4,j)**	**w(5,j)**
8	=SIN(PI()*G6)	=0
9	=G8*(1+$B5-0.5*$B5^2)	=0
10	=$D$4*F9+$D$5*G9-G8	=0

	I
6	**exact**
7	**u(0.4,t)**
8	=SIN(PI()*0.4)*(SIN(PI()*B8)+COS(PI()*B8))

	A	B	C	D	E	F	G	H	I
1	**Example**		**h** =	0.2					
2	**7.3.1**		**k** =	0.1					
3	**Hyperbolic**		**μ** =	0.5					
4	**PDE**		**μ^2** =	0.25					
5	k*π =	0.31	**2(1-μ^2)**=	1.5					
6		x =	0	0.2	0.4	0.6	0.8	1	**exact**
7	**j**	**t(j)**	**w(0,j)**	**w(1,j)**	**w(2,j)**	**w(3,j)**	**w(4,j)**	**w(5,j)**	**u(0.4,t)**
8	0	0	0	0.588	**0.951**	0.951	0.588	0	**0.951**
9	1	0.1	0	0.743	**1.203**	1.203	0.743	0	**1.198**

Screen 7.3.1 *Solution of the 1-dimensional wave equation*

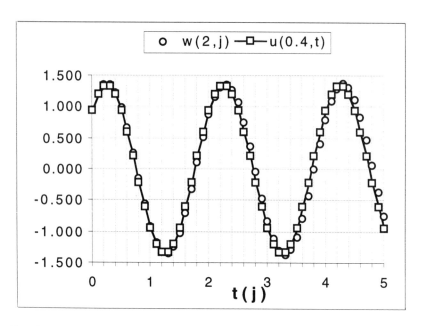

Graph 7.3.1 *Numerical and exact solutions of the wave equation at x = 0.4.*

Exercises 7.3

(1) Explore the behaviour of the numerical solution in Example 7.3.1 when k is given other values, for example 0.4, 0.2, 0.05.

(2) Alter the spreadsheet of Example 7.3.1 so that it uses either the approximation (7.3.3) or (7.3.5) instead of (7.3.4). Is the effect on the error as you would expect?

(3) Alter the spreadsheet of Example 7.3.1 to the case $M = 10$, and compare the accuracy with that found when $M = 5$.

Bibliography

ABRAMOWITZ, M., and I.A.STEGUN (1965), *Handbook of Mathematical Functions.* Dover Publications

BURDEN, R.L., and J.D. FAIRES (1989), *Numerical Analysis.* PWS-KENT Publishing Company

HENRICI, P. (1962), *Discrete variable methods in ordinary differential equations.* John Wiley & Sons

NORRIS, A.C. (1981),*Computational Chemistry, An Introduction to Numerical Methods.* John Wiley & Sons

SCHWARZ, H.R. (1989), *Numerical Analysis, A Comprehensive Introduction.* John Wiley & Sons

Answers to Selected Exercises

Exercise 3.4

The solution is $-\frac{1}{6}(5,4,3,2,1)$. The main changes in the spreadsheet can be seen in the following screen image.

	E	F	G	H
4	p(i)	q(i)	z(i)	x(i)
5	=B5	=C5/E5	=D5/E5	=G5-H6*F5
6	=B6-A6*F5	=C6/E6	=(D6-A6*G5)/E6	=G6-H7*F6
7	=B7-A7*F6	=C7/E7	=(D7-A7*G6)/E7	=G7-H8*F7
8	=B8-A8*F7	=C8/E8	=(D8-A8*G7)/E8	=G8-H9*F8
9	=B9-A9*F8		=(D9-A9*G8)/E9	=G9

Exercise 3.5

(1) For example, with right-hand side $(1,1,1)^T$ the three systems have solutions $(-3,7,9)$, $(0,-2,-3)$, $\frac{1}{9}(1,13,3)$ respectively. The Jacobi iteration is only convergent for the first of the three coefficient matrices, and the Gauss-Seidel iteration only for the third one.

(2) Part of the spreadsheet for the 5×5 system of Exercise 5.4 is shown below.

	A	B	C	D	E
1	Exercise 3.5	(2)/Ex3.4			
2	Jacobi	Jacobi	Jacobi	Jacobi	Jacobi
3	x(1)	x(2)	x(3)	x(4)	x(5)
4	-1	-1	-1	-1	-1
5	=(B4-1)/2	=(A4+C4)/2	=(B4+D4)/2	=(C4+E4)/2	=D4/2

	F	G	H	I	J
2	Gauss-Seidel	Gauss-Seidel	Gauss-Seidel	Gauss-Seidel	Gauss-Seidel
3	x(1)	x(2)	x(3)	x(4)	x(5)
4	-1	-1	-1	-1	-1
5	=(G4-1)/2	=(F5+H4)/2	=(G5+I4)/2	=(H5+J4)/2	=I5/2

Exercises 3.6.2

(2) Inclusion of relaxation enables the Jacobi iteration for the third coefficient matrix to be made convergent (with best $\omega \approx 0.31$), but not the second. For the Gauss-Seidel iterations, relaxation allows the first system to be made convergent (with best $\omega \approx 0.64$), but not the second.

Exercise 3.7

The spectral radii for the iteration matrices corresponding to the three coefficient matrices given in Exercises 3.5(1) are, respectively, 0, 1.77, $\frac{1}{2}\sqrt{5} \approx 1.118$ for the Jacobi iteration, and $2(1+\sqrt{2}) \approx 4.83$, 2, 0.5 for the Gauss-Seidel case.

Exercises 4.3.1

(1) The point to note is that $g(x)$ does not satisfy the conditions of theorem 4.3.1, yet it does have a unique fixed point in [4,5].

(2) The last three iterations are convergent, but not the first two. The root lies at $s \approx 1.426$, and the approximate values of $|g'(s)|$ for the five cases are, in order, 13.7, 2.6, 0.713, 0.161, and 2.4×10^{-16}. The last case

is an example of quadratic convergence (the iteration was derived using Newton's method) and in fact $|g'(s)| = 0$ here.

Exercise 4.3.3

The iteration cannot converge to the root $s = \frac{1}{2}[3 + 2a + \sqrt{1 + 4a}]$, because in this case $|g'(s)| > 1$.

Exercises 4.6

(1) The solutions are (-0.5187,-1.3078) and (1.6015,0.8540), to 4 decimal places.

Exercises 5.2

The error when the midpoint rule is used with $N = 20$ is (1) 0.000078, (2) 0.0017, (3) 0.026, (4) 0.000084, and $erf(1.5) = 0.9661051465$, to 10 decimal places.

Exercise 5.3

The values of N needed to achieve at least the same accuracy using Simpson's rule as when the midpoint rule with $N = 20$ are (1) $N = 3$, (2) $N = 2$, (3) $N = 4$, and (4) $N = 3$.

Exercises 5.4

(2) $N \geq 164$ will guarantee $|R_N| \leq 10^{-4}$ when the midpoint rule is used.

Exercises 5.5

(1) The integral has exact value $e^{-1} + \frac{\sqrt{\pi}}{2}(1 - erf(1)) \approx 0.50728223$. The transformed integral is

$$\int_0^1 t^{-\frac{5}{2}} e^{-\frac{1}{t}} dt$$

For this integral, the trapezoidal rule with $N = 20$ yields the estimate 0.50716, with error 0.024%, and Simpson's rule with $N = 5$ gives 0.50552, with error 0.35%. To achieve the accuracy of 0.1% or better requires that the trapezoidal rule be used with $N \geq 11$, or Simpson's rule with $N \geq 8$. Simpson's rule does not show its usual high accuracy because the fourth derivative of the integrand can be very large on [0,1].

Exercises 5.6

(2) One scheme for extracting all the required trapezoidal rule approximations from the one table is shown below. The rest of the Romberg table is easily completed with the FILL command.

	A	B
1	Exercise 5.6(2)	
2	Romberg	
3	Integration	
4		T(m,n)
5	m\\n	=1
6	=1	=D34
7	=A6+1	=E34
8	=A7+1	=F34
9	=A8+1	=G34
10	=A9+1	=H34

	C	D
1	a =	=0
2	b =	=1
3	smallest h =	=(D2-D1)/16
4		
5	=B5+1	
6		=C5+1
7	=(4^(C$5-1)*B7-B6)/(4^(C$5-1)-1)	
8	=(4^(C$5-1)*B8-B7)/(4^(C$5-1)-1)	=(4^(D$5-1)*C8-C7)/(4^(D$5-1)-1)
9	=(4^(C$5-1)*B9-B8)/(4^(C$5-1)-1)	=(4^(D$5-1)*C9-C8)/(4^(D$5-1)-1)
10	=(4^(C$5-1)*B10-B9)/(4^(C$5-1)-1)	=(4^(D$5-1)*C10-C9)/(4^(D$5-1)-1)

	A	B	C
12			m =
13			16/(2^(m-1)) =
14	Trapezoidal	Integrations	
15	k	x(k)	f(x)
16	=0	=D1	=B16*EXP(-(B16^2))
17	=A16+1	=B16+D$3	=B17*EXP(-(B17^2))

	A	B	C
31	=A30+1	=B30+D$3	=B31*EXP(-(B31^2))
32	=A31+1	=B31+D$3	=B32*EXP(-(B32^2))
33			
34			Trap. Int. =

	D	E
12	=1	=D12+1
13	=2^(5-D12)	=2^(5-E12)
14	Sum for	Sum for
15	T(1,1)	T(2,1)
16	=$C16/2	=$C16/2
17	=D16+$C17*IF(MOD($A17,D$13)=0,1,0)	=E16+$C17*IF(MOD($A17,E$13)=0,1,0)

	D	E
31	=D30+$C31*IF(MOD($A31,D$13)=0,1,0)	=E30+$C31*IF(MOD($A31,E$13)=0,1,0)
32	=D31+$C32*IF(MOD($A32,D$13)=0,1,0)/2	=E31+$C32*IF(MOD($A32,E$13)=0,1,0)/2
33		
34	=D13*$D3*D32	=E13*$D3*E32

	F	G
12	=E12+1	=F12+1
13	=2^(5-F12)	=2^(5-G12)
14	Sum for	Sum for
15	T(3,1)	T(4,1)
16	=$C16/2	=$C16/2
17	=F16+$C17*IF(MOD($A17,F$13)=0,1,0)	=G16+$C17*IF(MOD($A17,G$13)=0,1,0)

	F	G
31	=F30+$C31*IF(MOD($A31,F$13)=0,1,0)	=G30+$C31*IF(MOD($A31,G$13)=0,1,0)
32	=F31+$C32*IF(MOD($A32,F$13)=0,1,0)/2	=G31+$C32*IF(MOD($A32,G$13)=0,1,0)/2
33		
34	=F13*$D3*F32	=G13*$D3*G32

	H
12	=G12+1
13	=2^(5-H12)
14	Sum for
15	T(5,1)
16	=$C16/2
17	=H16+$C17*IF(MOD($A17,H$13)=0,1,0)

	H
31	=H30+$C31*IF(MOD($A31,H$13)=0,1,0)
32	=H31+$C32*IF(MOD($A32,H$13)=0,1,0)/2
33	
34	=H13*$D3*H32

The preceding screen images should be studied in conjunction with the one immediately below, which shows the entire spreadsheet with values displayed rather than formulas.

	A	B	C	D	E	F	G	H
1	Exercise 5.6(2)		a =	0				
2	Romberg		b =	1				
3	Integration		smallest h =	0.0625		exact =	0.3160602794	
4		T(m,n)						
5	m\\n	1	2	3	4	5		
6	1	0.18394						
7	2	0.28667	0.32091					
8	3	0.30888	0.31629	0.315978				
9	4	0.31428	0.31607	0.316059	0.31606072			
10	5	0.31561	0.31606	0.316060	0.31606028	0.3160602787		
11								
12			m =	1	2	3	4	5
13			16/(2^(m-1)) =	16	8	4	2	1
14	Trapezoidal Integrations			Sum for	Sum for	Sum for	Sum for	Sum for
15	k	x(k)	f(x)	T(1,1)	T(2,1)	T(3,1)	T(4,1)	T(5,1)
16	0	0	0	0	0	0	0	0
17	1	0.0625	0.06226	0	0	0	0	0.06226
18	2	0.125	0.12306	0	0	0	0.12306	0.18532
19	3	0.1875	0.18102	0	0	0	0.12306	0.36634
20	4	0.25	0.23485	0	0	0.23485	0.35792	0.60119
21	5	0.3125	0.28343	0	0	0.23485	0.35792	0.88462
22	6	0.375	0.32581	0	0	0.23485	0.68372	1.21043
23	7	0.4375	0.36129	0	0	0.23485	0.68372	1.57171
24	8	0.5	0.38940	0	0.38940	0.62425	1.07312	1.96111
25	9	0.5625	0.40993	0	0.38940	0.62425	1.07312	2.37104
26	10	0.625	0.42290	0	0.38940	0.62425	1.49602	2.79394
27	11	0.6875	0.42855	0	0.38940	0.62425	1.49602	3.22249
28	12	0.75	0.42734	0	0.38940	1.05159	1.92335	3.64982
29	13	0.8125	0.41988	0	0.38940	1.05159	1.92335	4.06970
30	14	0.875	0.40691	0	0.38940	1.05159	2.33027	4.47661
31	15	0.9375	0.38928	0	0.38940	1.05159	2.33027	4.86590
32	16	1	0.36788	0.18394	0.57334	1.23553	2.51421	5.04984
33								
34			Trap. Int. =	0.18394	0.28667	0.30888	0.31428	0.31561

Exercises 6.6

(3) The phase-plane plot y versus x, and the graphs of x and y versus t are shown below, for the parameter values given in the problem.

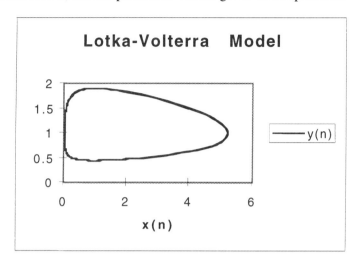

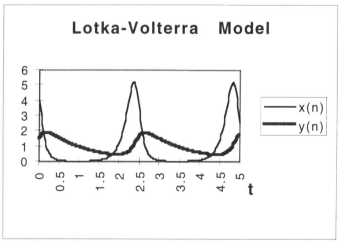

(4) The plots of sugar levels for non-diabetic, untreated diabetic, and twice-injected diabetic cases are shown below.

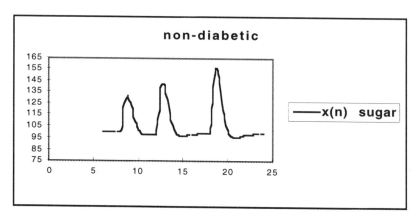

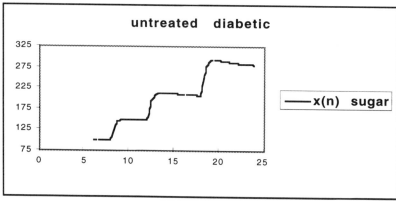

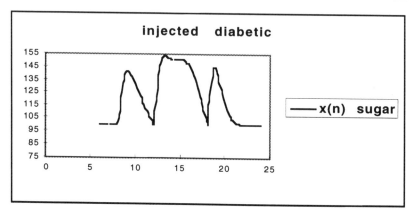

Exercises 6.7

(2) The graph of x versus t for the given parameter values is shown below,

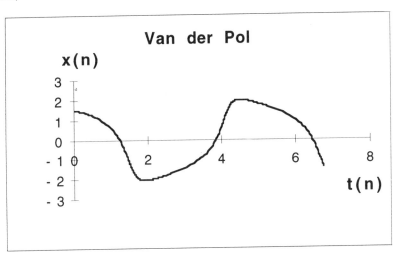

Exercises 6.10

(3) The variation in step-size is as below, for the given parameter values.

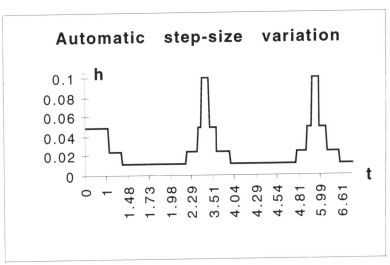

"Learning through Computer Algebra" Series
Published by Chartwell-Bratt. Series editor John Berry, University of Plymouth

Learning Mathematics through DERIVE
J. S. Berry, E. Graham, A. J. P. Watkins
This book develops foundation mathematics for scientists and engineers through the use of DERIVE. It is a standalone textbook, for post-16 through to undergraduate level, with DERIVE integrated as an investigative tool. DERIVE is used to introduce and help students to understand basic concepts in mathematics and as a problem-solving tool for solving real problems from the world of science and engineering. ISBN 0-86238-461-3, 370 pages, 1996

Learning Numerical Analysis through DERIVE
Terence Etchells, John Berry
This book covers the major numerical methods, and their analysis, for first courses at college and undergraduate level. The relative merits of each method are covered both analytically, providing a thorough grounding in the algebraic approach, and practically, through the tried and tested computer lab-based activities.
DERIVE provides a platform on which to quickly and accurately perform many complicated numerical calculations. Also, DERIVE's ability to algebraically manipulate expressions and perform calculus operations, enhances the investigation of the convergence of numerical methods.
Each chapter includes the development and algebraic analysis of the methods, lab-based activities, ideas for coursework, case studies, exercises and solutions. Free supporting utility files are downloadable via Chartwell-Bratt's web server.
Chapter 1 introduces the basic tool of numerical methods, which is recurrence relations, their solution and ill-conditioning problems. In chapter 2 we use recurrence relations methods that are used in solving equations. Chapter 3 deals with the approximation of functions by polynomials, and in particular the Taylor Polynomial, which is then used extensively in chapter 4 to analyse the errors associated with numerical methods.
Chapters 5 and 6 deal with numerical approaches to the calculus of differentiation and integration. In chapter 7 we introduce and analyse numerical methods of solving differential equations.
ISBN 0-86238-468-0, 239 Pages, 1997

Learning Linear Algebra through DERIVE
B. Denton, ISBN 0-86238-466-4, 296 pages, 1995

Learning Differential Equations through DERIVE
B. Lowe, J. S. Berry, Available Autumn 1997.

Learning Modelling with DERIVE
S. Townend, D. Pountney
ISBN 0-86238-467-2, 256 pages, 1995

Other books on using DERIVE in education

Mathematical Activities with DERIVE

E. Graham, J. S. Berry, A. J. P. Watkins (eds)
ISBN 0-86238-478-8, 216 pages, 1997, A4 spiral bound

Improving Mathematics Teaching with DERIVE: a guide for teachers

B. Kutzler, ISBN 0-86238-422-2, 185 pages, 1996

Elementary Linear Algebra with DERIVE: an integrated text

R. J. Hill, T. A. Keagy, ISBN 0-86238-403-6, 392 pages, 1995.

Other mathematics technology books (send for full catalogue)

Computer Algebra in Mathematics Education: State of the Art

J. Berry, M. Kronfellner, B. Kutzler, J. Monaghan
ISBN 0-86238-430-3, 168 pages, 1997.

Practical Mathematics using MATLAB

G. Bäckström, ISBN 0-86238-403-6, 241 pages, 1995.

Mathematics in Action:
Modelling the Real World Using Mathematics

Richard Beare

This lucid and readable introduction covers modelling as a process and includes an interest in the mathematics for its own sake. It is suitable for college, undergraduate and postgraduate study, whether of mathematics or other disciplines, since the wide range of models are treated at various levels.

A unique feature is that most models discussed are implemented in spreadsheet form on the associated CD-ROM. Interactively exploring their behaviour is a far more effective (and interesting) way of learning than merely reading the printed word and looking at static diagrams in a book. However, the book is also designed to be read on its own, without the use of a computer, since ample example results from all the models are shown in the illustrations. Realistic examples are used, which are of practical or theoretical importance to someone, and which exemplify both the process of modelling and links with conventional topics of mathematical study. For this reason, the chapters are arranged to correspond to particular types of mathematical representation. Consequently, the reader will find it easier to connect examples of mathematics in action with different aspects of mathematics that he or she is already familiar with, and that systematic development is possible avoiding the complexities and inelegance that so often result when one simply takes modelling problems and finds out where they lead.

The book also aims to make clear how the same real world situation can often be modelled using a variety of different mathematical approaches and how these link with one another.

Models have been chosen from as wide a range of subject disciplines as possible and for their intrinsic interest, but are readily understandable to those without specialist knowledge in the relevant fields. Well researched relevant background information and theory for each modelling problem is presented, so that it is placed in its proper context and its true significance appreciated. The models relevant to difference subject disciplines can be found by looking in the detailed index under headings such as geographical models, engineering models and so on. 531 pages, 1997, ISBN 0-86238-492-3

Mathematical Activities with Computer Algebra: a photocopiable resource book T. Etchells, M. Hunter, J. Monaghan, S. Pozzi, A. Rothery
Aimed at students in the over-16 age range in sixth forms, colleges and first year higher education in universities, in the UK and USA.
ISBN 0-86238-405-2, 126 pages, 1997.

Technology in Mathematics Teaching: a bridge between teaching and learning L. Burton, B. Jaworski
ISBN 0-86238-401-X, 496 pages, 1995.

Mathematics Handbook for Science and Engineering
Råde L - Westergren B
The latest version of the most comprehensive mathematics reference book available for scientists, engineers and university . As well as classical areas of maths such as algebra, geometry and analysis, it also covers areas of particular current interest: discrete mathematics, probability and statistics, programming and numerical statistics. It concentrates on definitions, results, formulas graphs and tables and emphasizes concepts and methods with applications in technology and science. ISBN 0-86238-406-0, 539 pages, 1995

Numerical Methods of Integration H. V. Smith
This book describes, with the aid of worked examples and supplementary problems, many of the more recent and important techniques for the numerical evaluation of definite integrals. Extremely useful to undergraduate/postgraduate students, engineers, mathematicians and scientists; in fact, to anyone who has to approximate a definite integral.
ISBN 0-86238-331-5, 147 pages, 1993.

For details of other mathematics books email
philip@chartwel.demon.co.uk